Secret Lives of Ants

Secret Lives of Ants

Jae Choe

foreword by JANE GOODALL

photographs by DAN PERLMAN

translated by DAN LEONARD

the johns hopkins university press

baltimore

This book has been brought to publication with the generous support of Korea Literature Translation Institute (KLTI) for the project "Books from Korea," 2005.

The Johns Hopkins University Press
2715 North Charles Street
Baltimore, Maryland 21218-4363
www.press.jhu.edu

Library of Congress Cataloging-in-Publication Data
Choe, Jae C.
  Secret lives of ants / Jae Choe ; foreword by Jane Goodall ;
photographs by Dan Perlman ; translated by Dan Leonard.
      p. cm.
  Includes bibliographical references.
  ISBN-13: 978-1-4214-0428-8 (hardcover : alk. paper)
  ISBN-10: 1-4214-0428-1 (hardcover : alk. paper)
  1. Ants—Behavior. 2. Social behavior in animals. I. Title.
  QL568.F7C55 2012
  595.79'6—dc23   2011021207

A catalog record for this book is available from the British Library.

Title page illustration: After her nuptial flight, the queen ant uses her middle and hind legs to break off her wings, as she prepares to start her new home.

Special discounts are available for bulk purchases of this book. For more information, please contact Special Sales at 410-516-6936 or specialsales @press.jhu.edu.

The Johns Hopkins University Press uses environmentally friendly book materials, including recycled text paper that is composed of at least 30 percent post-consumer waste, whenever possible.

. . . .

To my mother, a queen who gave birth to four males who never held a broom, who spent her whole life working like a worker ant

CONTENTS

FOREWORD

. . . .

I first met Jae Choe in 1996 when he interviewed me for a science magazine during my first visit to South Korea. I was immediately impressed by his warmth, his passion for the natural world, and his conviction that it is important to share scientific findings with the general public—a conviction that I share. When the interview was over he showed me an issue of *National Geographic* that, coincidentally, featured articles by each of us. We had a wonderful discussion about our work and our shared love for the natural world. Each time I've returned to Korea to lecture and speak with young people, Jae has acted as my liaison and interpreter. Of course, I cannot understand what he is saying when he translates, yet the reactions of people listening prove that he is able to convey not just my words but also—almost more importantly—the spirit of my message to audiences all over the country.

My own experience with ants is limited and very unscientific. There were wood ants that piled up small twigs and pine needles to form big, mound-shaped nests in the woods where I and my friends played when we were children. Their bite was painful and we used to avoid them. Then I decided that a friendly approach was necessary and trustingly held my finger toward one of them—whereupon she promptly clamped her jaws into my flesh!

In Gombe National Park, the site of my ongoing research, army, or driver, ants are commonly preyed upon by chimpanzees. The chimpanzees choose long, straight sticks and, by carefully peeling the bark and any projecting twigs, fashion smooth tools. These are pushed into an underground nest and withdrawn along with a mass of insects that have bitten on in defense of their nest. The chimpanzee pulls the stick through his free hand, rushes the ants into his mouth, and chews as fast as possible. He can endure this painful feast for only a few minutes before rushing off to pluck the biting ants from his body. I quickly learned that it is important to try to avoid the trails of army ants encountered in the forest. It seems to me that they have an unpleasant sense of humor since, if you do inadvertently walk through a trail, the ants are quick to climb your leg—but you are not aware until the leader reaches about knee level or higher, at which point the stream of ants that have climbed, as though in response to some signal, all bite at the same time!

Weaver ants taste hot and lemon-like and are another tasty chimpanzee snack. They make nests by sticking living leaves together while still attached to bush or tree. The chimpanzee picks the whole nest and, having crushed many of the ants by pulling it several times through hand or foot, then carefully opens the leaves and picks off the ants. Chimpanzees eat carpenter ants by poking short, thin sticks into their nests in dead wood and eating them from trunk or branch when they swarm out. This is a new behavior for the chimpanzees of our main study group, learned from a female who had acquired the skill in her natal community.

It is wonderful to watch the "clicker" ants that set out, in small raiding parties, to capture termites. The successful raiders then return to their nest, each with an unfortunate victim in her jaws. Infant chimpanzees love to poke sticks into trails of ants and watch with absorbed fascination as they scurry around seeking the source of the disturbance. My son, as a small child, did exactly the same.

It is his passion for the natural world that Jae brings to the study of ant biology. As he admits in *Secret Lives of Ants*, ants are not beautiful—by common human standards, at any rate. Chimpanzees, especially infants with their big luminous eyes, can touch the heart and serve as a flagship species in a campaign to save a rainforest, but it is extremely unlikely that zoomed-in close-ups of bristled ant faces, antennae splayed, could have the same effect. Sadly, most people are not even interested in ants and have no idea of their fascinating social structures and behavior. Yet ants are extraordinarily adaptable and successful and can be found in every environment except for the liquid, molten, or frozen places of our Earth. How fortunate that Jae is not only passionate about ants but equally passionate to rouse interest in the general public.

The result is this enchanting book. Jae draws simple analogies between human and ant societies, picking out aspects of our human societies such as agriculture, warfare, and the formation of alliances in order to help us understand similar patterns in ant societies. Ants, like humans, have developed sophisticated behaviors to help them deal with problems caused by living in crowded conditions similar to those we experience in cities. And as we read in these pages about some of the incredible achievements of ant communities we become more and more fascinated and amazed. Leaf-cutter ants, we learn, are farmers, harvesting mushrooms: their queens have been handing down knowledge of mushroom species to their daughters for more than 25 million years. Other ant species herd aphid livestock to provision their colony with a steady supply of food. Army ant colonies wage wars and have been observed to use signature war-plan tactics. And honeypot ants actually have individuals who call for reinforcements if a battle is not progressing well.

Certain ant species even create slave societies, tricking captured workers into laboring for a queen who is not their mother by exposing them to socializing pheromones. Some queens have been observed to ally with other queens—amazingly even those of a different ant species—in the ambitious and difficult task of founding a new colony. All ants live within highly structured and organized cooperatives. As in the case of human societies, some ant colonies are nomadic; others flourish in densely populated, multispecies communities within a single tree; some species are gatherers; others hunt. Some are quite primitive, and individuals within the society communicate only by touch, while others have a complex system of chemical fragrances that is the basis of what might be called a rudimentary language. Like early human explorers in bygone eras, many ants navigate by means of the angle of the sun and have internal clocks that adjust for its changing position in the sky over time.

But Jae wants this book to do more than help us to understand ants better. He suggests that we humans, our species a newcomer to this planet, could learn a great deal from the ants' highly successful 100 million years of experience in the field. For we have not managed our affairs well. We are a selfish species and as our numbers have multiplied we have placed increasingly unrealistic demands on the nonrenewable resources of the planet; we have polluted air, water, and land; there is unequal distribution of wealth; and today we face a horrific lack of understanding between extremist factions of different religions and a seeming inability to resolve the resulting crises. We became the masters of the planet by using our brains, but we have lost the wisdom that required decisions be made that would benefit and not harm future generations.

As Jae points out, ants have similarly evolved as the masters of their domain—yet without destroying their ecosystems. Indeed, comments Jae, an alien observer might even wonder whether ants or humans were the real masters of the planet. Despite crowding, they have managed to avoid high incidences of contagious disease. Though there is a clearly delineated division of labor in an ant colony, they have avoided overspecialization of roles, which, in human societies, as Jae points out, can cripple an organization's ability to adapt and innovate in times of crisis. We may see the ant colony as a "superorganism" within which every ant has a role to play. And, says Jae, though ants are individual actors with singular interests in genetic reproduction—much like the various specialized cells within our bodies—they carry out their tasks for the larger good of the colony, and in this regard theirs is the ultimate unselfish existence. While our individualist tendencies are part of what makes us special, Jae points out that we could certainly take a lesson from the ants in the creation of a smoother, more just, and more efficient society.

Ants are perhaps the most significant component of ecosystems everywhere: to understand them better is to understand nature better. And *Secret Lives of Ants* certainly helps us to understand and appreciate ants. It also helps us to appreciate that we are part of the whole, glorious tapestry of life on Earth. I share with Jae the hope that this book will foster a love of nature in young people and inspire them to study the wonders around them, whether they choose the largest mammals like elephants and whales or the smallest insects—like ants. For, of course, there is still much to learn about these amazing ant-people. Many questions are as yet unanswered, and even more are as yet unasked. And this despite the fact that as long ago as the reign of King Solomon, people were aware of the wonders of ant behavior: "Go to the ant thou sluggard, study her ways and be wise" (Proverbs 6:6).

The more we understand about the wondrous world in which we live, the greater will be our desire to protect and save its diverse habitats and the creatures that live in them. I congratulate my friend Jae on writing a truly fascinating book and I hope that you will buy a copy and give copies to your friends. And that it finds a place in school and university libraries so that it can be read by as many people as possible.

JANE GOODALL, PHD, DBE
Founder, The Jane Goodall Institute
& UN Messenger of Peace
www.janegoodall.org

PREFACE

I give a lot of lectures for the general public and schoolchildren, sometimes more than a dozen a month. My career as a public speaker began shortly after I returned to Korea from the United States in 1994. I even appeared on national TV and gave a six-month-long weekly lecture series on animal behavior. People seemed to enjoy listening to my animal stories but rarely asked me questions afterward. It was different, however, when I gave a lecture on ants. Quite a few hands went up before I even finished talking. There were questions like, "Can we talk with ants?" and "Can ants learn new techniques and improve their lives?" "What imagination!" I thought; I'd been studying ants for more than a decade but had never asked such imaginative questions myself. I was so impressed by these questions that I asked the audience how they could come up with them. They all told me about the science-fiction novel entitled *Les Fourmis* (*Empire of the Ants*) by French author Bernard Werber.

I rushed to a bookstore, bought the book, and read it overnight. It was quite interesting indeed. But soon, such questions started to rather annoy me, because a nearly identical set of questions hounded me every time I gave an ant lecture. As interesting as it was, Werber's novel was fiction. So, I decided to write a book of my own based on scientific facts. Mark Twain once commented on truth being "stranger than fiction"; indeed, the reason that Werber's novel is so impressive and does not seem to be as hard to swallow as many other works of science fiction is that Werber himself had spent many hours observing ants since he was a child. Nevertheless, the amazing reality of ant society is many times more fascinating than anything even Werber was able to imagine. In fact, my book, too, merely scratches the surface. I hope that many of the students who read this book go on to study ants and many other mysteries of nature as well.

In writing this book I have tried to use a conversational style and ordinary language to describe the results of a great deal of scientific research on ants. In 1990 Bert Hölldobler and Edward O. Wilson of Harvard University compiled and published the seminal work *The Ants*—732 pages measuring 26 x 31 centimeters (10.2 x 12.2 inches), weighing in at 3.4 kilograms (7.5 pounds). The well-known scientist and author of *The Selfish Gene,* Richard Dawkins, once said that it does no good to cheat readers in the name of writing simply. This means that accuracy must not be sacrificed in this process. Despite my best effort to achieve this, I am sure that there must be parts that are not so easy to understand, but I do hope that you the reader will see them as opportunities to challenge yourself.

In this book I tell stories of my own journey to the world of the ants. I also describe the economic structure, society and culture, and political systems of the ant world. By looking at these creatures who, long before human beings, developed ranching, dairy farming, agriculture—a highly developed division of labor that resembles the assembly lines of humankind's automobile factories—and multinational enterprises we get a glimpse of ourselves. I have also included tales of ant culture, its foundation of self-sacrifice and a finely tuned chemical language, as well as how this culture has helped to maintain systems of monarchy and slavery. Ants are among the few creatures besides humans to engage in large-scale warfare, and the tales of their massacres and atrocities, as well as the struggles for power in ant monarchies, are all too reminiscent of our own. In some respects, ants are more like us than is the chimpanzee, whose DNA is nearly 99% identical to our own. Although physically ants do not resemble us at all, if I had to choose the animal whose way of life most closely resembles modern mankind, I would not hesitate to choose the ant. By observing the way ants live, we can clearly see how strikingly similar their lives are to our own. At times my anthropomorphic imputation of human qualities to the ants may be pushed a bit too far. I hope that readers will understand that I am using these comparisons not in a literal sense but as a heuristic device to help them appreciate the marvels of ant behavior.

A great deal of the material in this book has been published in an earlier form serially in a popular Korean science magazine, *Dong-a Science.* I owe the editors and reporters of *Dong-a Science,* especially Dr. Kim Hoon-gi, a huge debt of gratitude. I also want to thank Lee Gap-su, Lee Choong-mi, and the rest of the editing staff of Science Books for their hard work in publishing quality books about science. It is unusual for a textbook on Korean literature to include an essay by a scientist, but what appears here as chapter 7 was included in the Korean Literature textbook for middle schoolers. This was done as part of the recent effort to bridge literature and science in educating middle school and high school students in Korea. The publication of this English version was possible only because the Korea Literature Translation Institute selected my book as part of its promotion to introduce Korean literature to the global community. The institute provided financial support to hire a native speaker for help with translation and to defray the production costs.

I thank Dan Leonard for his help with making the book readable. Nonetheless, this book is not a straightforward result of translation: I again revised the translated manuscript by omitting some parts that were too provincial to Korean situations and adding information available from more recent discoveries. I am grateful to Vince Burke, senior editor at the Johns Hopkins University Press, who shepherded my manuscript through to publication, and Dennis Marshall, the copy editor, for elevating the readability of my book to another level.

The single item that makes the book stand out so beautifully is its many color photographs. Most of the photographs were taken by Dan Perlman of Brandeis University, my colleague and classmate from Harvard who researched ants with me for a long time. All but a handful of them were taken by him. Dan also read an earlier draft of the English version and gave me many critically important suggestions. Thank you, Dan. I would also like to thank

Dr. Hangkyo Lim of the University of Minnesota, who provided me with photos of the Japanese carpenter ant. Many of the ant drawings in this book have been redrawn from those of other myrmecologists and illustrators, including Katherine Brown-Wing, Helena Curtis, John D. Dawson, Turid Forsyth, Bert Hölldobler, Ulrich Maschwitz, Carl Rettenmeyer, J. Szabo-Patay, and E. O. Wilson.

Finally, I would like to extend my deepest thanks and love to my wife and son for being so forgiving and patient as I spent so many nights and weekends hunched over my computer, writing this book.

INTRODUCTION

. . . .

my first journey to the ants

In 1970, Soviet dissident author Aleksandr Solzhenitsyn won the Nobel Prize for Literature. I was in high school then and dreamed of becoming a poet or novelist someday. As soon as his collected works came out, I immersed myself in the literary world of Solzhenitsyn. Boris Pasternak, another Russian Nobel laureate from 12 years earlier, may be better known because he inspired the movie *Dr. Zhivago,* but Solzhenitsyn's writings about life in the Stalinist gulags, such as his novel *One Day in the Life of Ivan Denisovich* and his play *The Love-Girl and the Innocent,* made a particularly strong impression on me. Of Solzhenitsyn's writings, another that made a subtle but lasting impression on me was the short essay "The Bonfire and the Ants."

> I threw a rotten log onto the fire without noticing that it was alive with ants. The log began to crackle, the ants came tumbling out and scurried around in desperation. They ran along the top and writhed as they were scorched by the flames. I gripped the log and rolled it to one side. Many of the ants then managed to escape onto the sand or the pine needles.
>
> But, strangely enough, they did not run away from the fire. They had no sooner overcome their terror than they turned, circled, and some kind of force drew them back to their forsaken homeland. There were many who climbed back onto the burning log, ran about on it and perished there.*

The literary merit of this essay might not be the greatest, but for me, reading Solzhenitsyn's words was like throwing a log on the fire of scientific curiosity that was already starting to burn in my heart. Although that essay had even-

*Translation by Michael Denny. © Alexandre Soljénitsyne; used by permission.

tually led me to study science, I did not begin to unravel the fundamental mystery of ant behavior until many years later. In the summer of 1984 I was taking a course in tropical biology at the Organization for Tropical Studies (OTS) in Costa Rica. It was a field course for which approximately twenty graduate students majoring in biology were selected to tour and study the tropical habitats of Costa Rica. I was excited from the moment I set foot in that jungle scene that could have come straight out of the Tarzan movies I watched as a child. After everyone arrived in the capital city of San José, our first stop was the field research station in La Selva, a classic tropical rainforest habitat. I will never forget how my body tingled with excitement to finally experience for myself the tropical rainforest that I had only heard about.

After I stepped into the forest, I lost track of time. The canopy was so dense that it was dark even in the middle of the day. Suddenly, I caught a glimpse of tiny leaves moving. I froze in my tracks on the narrow trail. Those were not just leaves swaying in the breeze, it was a real leaf-cutter ant parade. Before this day, I had only seen leaf-cutter ants in the glass cages in E. O. Wilson's laboratory at Harvard. Watching the leaf-cutter ants marching in the tropical forest was a truly magnificent sight to behold.

A great deal of research has been done on these ants due to their unusual habit of carrying leaves back to their nests to use as a cultivation medium for growing mushrooms. They have earned the name leaf-cutter ants because they go into the forest to get leaves just like woodcutters go into the forest to cut down trees to get a supply of wood. They use their heavy, serrated jaws like saws to cut the leaves they then take back to the nest.

I followed the leaf-cutter ant parade until the sun went down, and after I returned to the research station I began to wonder if the ants continued to march in the dark. After I finished picking at the unfamiliar food of that unfamiliar country in the research station cafeteria, I strapped a light to my head and headed back into the forest. The area was full of all kinds of poisonous snakes and spiders, and even jaguars, but my curiosity was greater than my fear, so I forged ahead. Surprisingly, when I found the ants again

introduction

. . . . xxi

in the pitch-black jungle night, they were not carrying leaves; they were carrying pink flower petals. This beautiful parade of flower petals was simply mesmerizing.

I was excited by the thought that I might be the first person to observe leaf-cutter ants harvesting flower petals instead of leaves. I asked the other tropical biologists at the station, but no one could give me a definitive answer to my question. In those days the field research station did not have any research guides on ants, and there was no such thing as the Internet, so I had no choice but to wait until we returned to San José and I could go to the library. For the rest of my stay in La Selva, while going into the forest with the rest of the group by day to study many different kinds of natural phenomena—observing leaf-cutter ants hauling leaves in every spare moment I could find—after dinner I would take a cat-nap on my hard wooden bench and run right back into the forest to observe the ants.

When we returned to San José, I went to the library the first chance I had, but they did not have books that could answer my question. I had to wait another two months, when I returned to

A leaf-cutter ant parade carrying flower petals. © Jae Choe

Harvard, to find the answer. Two days after reaching Harvard I went to the library of the Museum of Comparative Zoology. My two months of excitement turned to disappointment when I found a short passage in an old natural history book that confirmed that decades earlier a European scientist had discovered leaf-cutter ants carrying flower petals. I had missed my imagined chance to shock the world with an amazing discovery, but I will never forget those nights I stayed awake following that trail of flower petals.

On the second expedition of my OTS course we went to the picturesque mountains of Monteverde, to the northwest of Costa Rica's Central Valley. It often rained, but the cloudy weather and moisture are what sustain that lush cloud forest ecosystem. Many Americans and Europeans settled there a long time ago because of the stunning beauty of the area, and it has been well preserved so that the rare and unusual birds of the area, such as quetzals and hummingbirds, can continue to lay eggs in their natural environment. It was also the last bastion of the Monteverde golden toad—now believed extinct. It was not that long ago, only the mid-1980s, but it was not so difficult to find groups of ten or more of the dazzling orange toads that could have hopped right out of a fairy tale. When I think about the fact that those amazing creatures may be gone forever, I cannot help but think that the world is an emptier place.

The night I arrived, I met Jack Longino, who was classifying the ants of Monteverde as part of his PhD research at the University of Texas. Knowing I was from Harvard, the Mecca of ant research, he invited me to look inside the trumpet trees where the Aztec ants lived. The next day I found a small trumpet tree on my way to the mountains. Trumpet trees, like bamboo, are hollow, and I found an amazing world of ants inside of this tree. In ant society, a single queen usually hatches worker ants by herself to found her colony. Many of the colonies inside this young trumpet tree, however, were founded by multiple queens working together. Even more amazingly, some of the queen ants who were cooperating together to form their colonies were from completely different species. Queen ants from the same species cooperating to found a colony was a rare enough occurrence, but queen ants from different species coming together to found a colony was a totally new finding.

At the time, I had already decided to write my doctoral thesis on the behavior and ecology of the insects called zorapterans—creatures evolutionarily similar to termites—but this amazing discovery left me in a serious quandary. When I returned to Harvard at the end of that summer I met Dan Perlman, a fellow Harvard graduate student who joined our laboratory to do his PhD one year after I had joined, and I encouraged him to research the Aztec ants. The next summer, Dan went to Costa Rica, as I had, to take the OTS course in tropical biology. When he returned after his 10-week course, his face was beaming with hope and excitement about his research on the Aztec ants. I was more than happy to turn the research topic over to him, but a part of me could not help but feel a little sorry for myself for passing up the chance to do the research. As if he could read my mind, Dan considered me his research partner from then on. Even though I did my thesis on zorapterans and he did his on Aztec ants, to this day he still considers me his fellow ant biologist. He is now a professor of environmental studies at Brandeis University.

Bert Hölldobler and E. O. Wilson, my academic advisors at Harvard University, are generally considered the world's foremost authorities on ants. For many years these two scientists collaborated on many research projects working out of their conjoined laboratory on the fourth floor of Harvard University's Museum of Comparative Zoology Laboratories. In the late 1980s, they decided to compile the twenty years of research they did together into a single, comprehensive book about ants. The result was the encyclopedia-sized text *The Ants,* which was published in 1990 and won a Pulitzer Prize two years later. *The Ants* was too lengthy and technical for the layman, so they published a shorter work, *Journey to the Ants,* in 1994.*

---

*Bert Hölldobler and Edward O. Wilson, *The Ants* (Cambridge, Mass.: Harvard University Press, 1990); Bert Hölldobler and Edward O. Wilson, *Journey to the Ants* (Cambridge, Mass.: Harvard University Press, 1994).

Aztec ant queens in the nodes of a trumpet tree. They found separate queendoms, but to be more competitive do not hesitate to form alliances with other ant queens, even queens from other ant species.

After completing my masters degree in ecology at Pennsylvania State University in 1982, the next year I began working on my doctoral degree in biology at Harvard. Unlike most doctoral candidates, when I began the program I already had a research topic in mind for my doctoral dissertation. More precisely, I was able to go to Harvard in the first place because Ed Wilson liked my research topic. At the time there were many well-supported theories about the origin and evolution of the social behavior of the Hymenopteran order of insects (which includes ants, bees, and wasps), but little research had been done on the evolution of their distant relatives, termites; consequently, I decided to begin my career in research at Harvard by turning my attention to termites, which in terms of behavior seemed so similar to other social insects, even though taxonomically they were so very different. However, working in the same laboratories with and regularly meeting the greatest authorities in the world on myrmecology (ant science), it was almost impossible for me not to get drawn into the study of ants as well. As a result, I ended up carrying out two projects, one on ants, and another on the Zoraptera, which at the time was believed to be the most closely related order of insects to termites.

After I was awarded my PhD in 1990, Harvard hired me as a full-time lecturer, and for the next two years I taught courses such as "Ecology," "Animal Behavior," "Social Insects," and "Human Behavioral Biology" for Harvard's biology and anthropology departments. After teaching "Animal Behavior" as a visiting assistant professor for the biology department of Tufts University, which is fairly close to Harvard, I moved on that summer to the University of Michigan's biology department as an assistant professor and a junior fellow at the Michigan Society of Fellows. In 1994 I took a position in the biology department of Seoul National University, where I continued my study of ants. I decided to begin my research of ants alongside the graduate students of Seoul National University by starting with the Japanese carpenter ants and their symbiotic relationship with the cherry trees on the Seoul National University campus.

It is easy to come into contact with ants because they are all around us. They also make ideal laboratory animals because they are relatively easy to rear in captivity. Even in a relatively small country like Korea there are well over 100 different species of ants. While it obviously depends on the research methods used, myrmecology is a field where a good idea and a strong pair of legs can be enough to accomplish world-class results. There are even high school students who are so engrossed with ants that they come to my laboratory day and night to study ants. It is my hope that many young myrmecologists will join me in discovering many more of the secret lives of ants that are waiting to be told.

PART I

The Economics of Ant Society

1

Ants Mean Business

how the futuristic economics of ants
maximizes their returns

Perhaps the best-known story about ants is "The Grasshopper and the Ant." A grasshopper fritters away the hot summer in the shade of the grass, having fun singing and eating the plentiful food around him to his heart's content. Meanwhile, the ant is toiling away ceaselessly in the hot sun, getting ready for winter. When the cold, snowy winter comes, the grasshopper finds himself going to the ant that he had made fun of all summer, to beg at her door for food.

However, this could only happen in the world of fairytales and the like. The grasshopper's song that comes from the grass all summer long is not a happy song of leisure: it is actually the sound of a male grasshopper's desperate struggle to attract a mate so that he can pass his genes on to the next generation. He has to sing louder and longer than the other males so that the female will notice him more easily. Even when food is plentiful, male grasshoppers have much to occupy them.

The Ants' Life of Hoarding

"The Grasshopper and the Ant" teaches us the value of working hard to prepare for hard times, but as a behavioral biologist I gleaned different lessons from the story. In the tale we see the grasshopper knocking on the ant's door on a cold snowy night, but in reality grasshoppers die off before winter even begins. The female grasshopper must hurry to lay her eggs deep underground and bury them before the cold sets in. When spring comes, the cycle of life continues

with the next generation of males hatching to face their own har-
rowing struggle to be the first to get their sperm into a female.

Since ants must live through the winter as adults, they are not
satisfied to just enjoy the food and water around them during the
season of plenty: they have to harvest and store the surplus for
the season of scarcity. In nature, aside from humans, squirrels,
ants, termites, bees, and other social animals, it is not so common
for animals to accumulate food beyond their immediate needs.
Of course, some animals forage throughout the winter, but most
animals, insects in particular, lay their eggs and die before winter,
like grasshoppers do. Otherwise, like migratory birds, for example, they travel great distances to find places where food is more
plentiful, or like bears they sleep through the winter.

Just how did ants get started on storing food? Humans, as hunter-
gatherers, may have stored some of the excess fruit they picked
or meat from their kills for emergencies, but full-scale food stor-
age for humans most likely did not begin until the onset of agriculture. Only after the industrial revolution did our economies
become complex enough for us to begin storing a wide variety of
resources other than food. But by the onset of mankind's agricultural age, ants had already been farming for a long time. The leaf-
cutter ants of the Latin American rainforests, who haul leaves in
their jaws in order to grow mushrooms on them, are considered to
have built the first farms on Earth as early as 50 million years ago.
In addition to ants that grow their own food, there are also ants
that herd aphids as humans might ranch dairy cattle. There are
even ants that have formed bilateral trade agreements with plants
to protect them from herbivores in return for food and shelter.

Specialists in a Single Job

After ants eat their fill, what makes them decide that it is better to
prepare for an unpredictable future than to kick back and relax in
the shade? A way to approach this question of how ants got started
on storing is to look at what they do now. "It's the economy, stu-
pid!" was an effective slogan used during Bill Clinton's successful 1992 presidential campaign against George H. W. Bush, and

the lives of ants revolve around a similar point of view. Economy, nothing but economy. Ants have established highly specialized divisions of labor in order to increase the productivity of their colonies.

Sociologically, the most remarkable system that ants have established is known as "reproductive division of labor." In this system, the queen ant is completely preoccupied with laying eggs, while the worker ants are in charge of taking care of everything that the queen and the colony need for reproduction. From the perspective of Darwinian evolution, in which the ultimate purpose of life is to pass one's genes on to the next generation, there are few greater mysteries than that the worker ants have given up on having and raising their own offspring to devote their whole lives to assisting their queen. While human beings are social animals, we have not systematically determined who can and cannot have children. Although the research of my colleague and an acclaimed evolutionary anthropologist Laura Betzig may show that in history, the ruling classes have had a much higher degree of reproductive success than the ruled classes, human society has not specialized the reproductive process. A few decades ago, Prime Minister Lee Kwan Yew of Singapore tried to establish laws to limit the number of children individuals could have according to their intelligence and level of social success, but most such attempts at eugenics have met harsh public criticism. But ants have their "reproductive division of labor."

The division of labor that the sexually segregated worker ants have established in order to maximize profits is reminiscent of the system used to organize modern-day factories. I am referring, of course, to the assembly-line system that American industrialist Henry Ford posited would maximize the profits of his automobile factories and that is still in use around the world. Highly organized worker ants are assigned specialized tasks. How tasks are assigned varies according to the species. In most ant societies, the worker ants all have the same body shape and size, but some ant species have worker ants that are divided from the day they are born into two or more classes. For worker ants with a uniform body shape

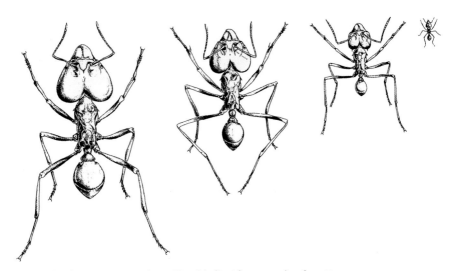

Leaf-cutter ants, planet Earth's first farmers. Leaf-cutters are divided into four classes from the day they are hatched.

and size, the work they are responsible for is not decided from the day they are hatched. Instead, they do different types of work as they grow older, tasks being assigned according to a system of seniority.

When these ants are young they serve the queen as nurses or midwives, looking after the eggs and larvae. When they reach, say, adolescence, they serve as nannies, feeding and bathing the larvae. When much older, they leave the underground tunnels to forage for food, defend the colony, or help with building new tunnels. At first they perform safer domestic chores, but as they grow older they participate in more dangerous work outside, such as defense or various kinds of construction and engineering.

Tight Control of the Division of Labor

Worker ants do not perform the exact same job day in and day out. Their age merely determines their general duties. When they are very young, they stay by the queen's side doing odd jobs, but when a little older they run around the nest doing all kinds of chores until they are old enough to work outside the nest. In ant societies with more than two classes of worker ant, the division of labor becomes more specialized. In general, the larger the worker ant's body is, the more specialized her job. In the most extreme cases the smaller worker ants do nearly thirty different types of work, while the larger workers will be involved only with a few kinds of work. In my laboratory research involving Japanese carpenter ants (*Camponotus japonicus* Mayr), which have two classes of worker ants, it was clear that the larger workers did more specialized work than the smaller ones.

In some species of ants, larger workers do only one kind of job. Fire ants, an agricultural pest in the southern United States, have large workers whose only job is to use their millstone-like jaws to grind large seeds. There are also ants that have classes of Goliath-like worker ants that leave the nest only to ruthlessly crush the bodies of enemies during wars with other colonies. Turtle ant col-

Large and small leaf-cutter worker ants. The large worker ant crushes the colony's enemies with her enormous jaws.

onies have a separate class of workers for guard detail. These ants' sole duty in life is to protect the nest by blocking the entrance with their unusually wide, flat heads.

Flexible Systems of Business Management

The advantages of the division of labor according to age or body type are obvious. Animals that give birth to and raise their young alone must constantly alternate between caring for themselves

Top, A turtle ant with a flat head. *Bottom*, Turtle ants use their flat heads to block the entrances to their nest.

and caring for their young. Ant colonies are as intricate and incessantly productive as our modern factories. Most ant factories are organized to manage many different processes simultaneously so that if a problem occurs with one process it is not a major setback for the enterprise as a whole. Such parallel processing has many advantages over serial processing, which is the default option for most solitary animals.

Economists predict that it is small, flexible businesses that will be the best-equipped to face the volatile market conditions of the twenty-first century. In other words, while nothing can surpass highly organized divisions of labor in terms of maximizing productivity, overly rigid business models are no longer believed to translate to successful business management. Harvard's Edward O. Wilson performed experiments to learn more about the flexibility of worker ants in the highly stratified leaf-cutter ant colonies. He found that when the number of workers in one class decreased, workers from other classes would pitch in to make up for the labor shortage in that area. However, if the adjustment was too extreme, such as large warrior or tunnel-building ants taking on nursery or maid duties, they would do their new jobs poorly in spite of their best efforts. By comparison, species with a single class of worker ant can assist with or switch jobs much more easily. Ants have a much longer history of industrial economy than humans do, and they have had time to adjust. This is why most ant societies contain only a single class of worker ants, even though specialization is clearly more advantageous. I believe we can learn a few things from the ants' theories of business management.

2

Economies of Scale and Rational Business Management

. . . .

from joint ventures to multinational enterprises

People frequently compare ant colonies to factories because ants have developed elaborate divisions of labor in their colonies in order to maximize their profits, much like human industry has. However, unlike our companies, some of which go bankrupt at the slightest downturn in business conditions, ants are able to reinvest most of their profits to expand their facilities and build new factories. Industrial economists, specialists in a subset of microeconomics, examine companies' investments of resources such as land, labor, and capital and how they use them to produce products and services. They also examine the decisions companies must make about how to maintain the balance between supply and demand, as well as which resources to invest in, and to what degree, to produce the right amount of product.

Worker Ants Are Nothing but Investment

The basis of the economics of ant factories is not different from that of human factories. Looking at ant colonies from an industrial economist's point of view, the nest is territory or factory, the number of worker ants is the labor force, and a colony's capital is the food the ants have gathered and stored. Ultimately, the products they are investing these resources into manufacturing are new generations of queen ants and reproductive males. Just as the companies who manufacture and sell the most products will control the greatest share of the market, the ant colonies that produce more queens and males than the colonies around them will likely be the most successful.

Even though worker ants are living, breathing organisms, from a strictly industrial economic point of view, they are essentially investments of capital to produce more queens and males. Worker ants serve the same function as the somatic cells that make up the bodies of living organisms, which are necessary to produce eggs and sperm, the reproductive cells. This is why ant colonies are sometimes called "superorganisms." Even though they are collectives of many living, breathing organisms, each colony is organized for a single purpose, as is the body of a single multicellular organism. Viewed as a single organism, the queens and males function as reproductive organs, while the workers function as the many parts and organs of the body.

When comparing ant economics and human economics, one of the most interesting phenomena in the ant world is related to one of the fundamental principles of economics—economies of scale. Just as companies that are the right size grow faster and are more efficient than companies that are too large or too small, research has shown that, in most cases, the most efficiently organized ant colonies are medium-sized. Small ant colonies frequently fail during their formative stage, just as newly founded small businesses frequently go bankrupt. Likewise, when ant colonies grow too large, communication breaks down and they can no longer react effectively to changes in their environment. These are just some of the many similarities we can see between ant society and human society.

The Nuptial Flight

The ant colony begins with the nuptial flight of virgin queens and males. On this special day, when the weather is warm and tranquil, the worker ants open the door to the nest and the maiden queens emerge. The queens meet the bachelor males from other colonies and hold their wedding ceremonies at the traditional rendezvous where generations of their grandmothers met their grooms—at the top of a hill or tall tree. After consummating her marriage, the wannabe queen breaks off her wings, which she will never need again, and finds a good spot to make her new home. Different

After her nuptial flight, the queen ant uses her middle and hind legs to break off her wings as she prepares to start her new home.

species have different nesting habits, but most queens build their nests underground, in rotten trees or inside plant stems.

The queens' and males' nuptial flight is the tensest and most exciting time in the ant calendar. On the day of the nuptial flight, the workers busily stream in and out of the nest while the maiden queens and bachelor males pace back and forth at the door like marathon runners waiting for the start of a race. I have watched the queens and males of the Japanese carpenter ants on the Seoul National University campus gathered near the entryways of their nest for days on end. Then one day, many of them come out, running this way and that but not yet flying off. The worker ants hold up their antennae, as if trying to detect something in the air. Ant colonies sometimes experience severe famine when food stocks run out, when natural disasters like flood or landslide strike, or when there are heavy casualties during war, but even then the ants strive to bring the colony back to normal without missing a beat. If they miss the nuptial flight for any reason, however, all their hard work for an entire year has been for nothing. It is as if they worked day and night manufacturing a product, but never took any of it out of the warehouse. Companies must make enough products to meet the demands of peak season, and being on time for the nuptial flight is no less crucial for an ant colony. They have to be ready to hold the wedding ceremony so the bride can meet her grooms from the neighboring colonies.

From a Closed Economy to an Open Economy

Most, if not all, queen ants start their new families shut away in a small chamber, not eating, completely absorbed in giving birth to and raising her offspring. It is too dangerous for her to make unnecessary trips outside for food, so to gain energy while she raises

Brachymyrmex male ants awaiting the nuptial flight.

her first worker ants she breaks down the fat stored in her body and also feeds on her now useless wing muscles. She must build a strong enough army to protect her before her limited resources run out. If she cannot produce enough workers to bring food from the outside in time, her queendom will fall before it has even had a chance to get off the ground. She must win this race against time to make the colony completely self-sustaining.

An Aztec queen making her new home inside the stem of a trumpet tree. Small, oval-shaped eggs are visible; also large and small larvae.

Sister wasps working together to found a new colony. The superior female will become queen; the others help establish the colony.

In economic terms, the queen hatching and raising her first worker ants, closed off from the outside world, could be called a closed economy. The key to success in an environment with limited capital is to manufacture the product as quickly and efficiently as possible. One way the queen can increase her capital or reduce her production time is through a joint venture with other queens. In these alliances, several queens begin their new colony together in the same chamber, cooperating throughout the entire process of founding the new colony, from laying and hatching the eggs to taking care of the larvae. These alliances frequently occur in areas where resources are limited and competition is particularly severe.

Many social wasps also form new colonies with multiple females. According to current research, however, female wasps will form new colonies only with females who are very closely related to them—sisters, in most cases. Eventually, one of the females will ascend to the throne as queen, while the others become workers.

Amazingly, ants do the exact opposite. Recent field research and molecular genetic analysis of DNA sequences have revealed that the queen ants who form colonies together are almost always, genetically, complete strangers to each other. The alliances in the ant world are believed to stem from economics, not genetics.

In the recently published *Encyclopedia of Animal Behavior* (2010), I was an editor for the arthropod social-behavior section and contributed a chapter titled "Colony Founding in Social Insects." In my review of literature, I came up with the following five potential advantages for queen associations in ants: (1) faster production of much larger first workers; (2) increased survival of cooperating queens; (3) earlier maturation to the reproductive stage; (4) better protection from parasites and predators; and (5) reduction of the costs of nest construction and maintenance. These joint ventures usually occur only within the same species, but there are interesting cases of queens from different ant species forming alliances. This has been observed in Aztec ants in the tropical rainforests of Costa Rica—a practice not unlike that of companies that increase their opportunities by rising above national boundaries to form multinational corporations.

Once the worker ants dig their way out of the nest to start bringing resources in from the outside world, the ant colony changes from a closed economy to an open economy. From this point on, the colony's success or failure will be determined by the queen's investments in workers and how efficiently these workers do their work. The colony moves beyond the stage of managing limited resources internally to reach a new stage where it must become internationally competitive. Modern business management schools study research on "business restructuring" to help businesses find their most efficient size and method of organization, but ants solved this problem long ago through a lengthy process of trial and error. As today's leading companies restructure on a large scale, the question of how ants found their ideal size and structure would be a fascinating topic for future research.

3

A 50-Million-Year Tradition
of Farming

the massive underground mushroom farms
of the leaf-cutter ants

It has been said that farmers are the founders of human civilization, but mankind's first farms are believed to be no more than ten thousand years old. This so-called agricultural revolution and the subsequent industrial revolution made us the mightiest species on Earth. Although in numerical terms we humans were not so superior during our hunter-gatherer years, since the advent of agricultural society and the industrial revolution our numbers have grown at such a staggering rate that we are now concerned about overpopulation. Just how did mankind begin this astonishing activity known as agriculture? How did people come up with the idea of taking seeds from the grass and fruits they found strewn around the mountains and plains and planting them in the ground so they could have more to eat later? Human beings are truly amazing creatures.

The First Agricultural Society

According to research done by the entomologists, animals had already begun farming at least fifty million years before the dawn of human agriculture. Leaf-cutter ants, which inhabit the entire tropical region of the Americas, were actually the first farmers on Earth. If you visit the tropical rainforests of Latin America, you can see long parades of leaf-cutter ants carrying in their jaws leaves that are often larger than their entire bodies. These leaf-cutter parades, stretching sometimes over up to several hundred meters, are magnificent spectacles to behold. I will never forget the feeling I had in 1984 when I saw a leaf-cutter ant parade for the first time.

17

I was walking on a trail in the La Selva tropical rainforest when I saw a lot of little green things move in lines. For as far as the eye could see, a parade of leaves both large and small was streaming through the forest. Seen from a distance there was no sign of the tiny ants, only their leaves and the sunlight that splashed upon them as they flowed by.

These ants do not harvest leaves for their dinner: they are actually gathering them to use as a cultivation medium for a unique breed of mushrooms that only leaf-cutter ants grow. The ants provide the mushrooms with more plant matter to eat than they would ever have on their own, and the mushrooms, growing quickly, in turn provide the ants with fruiting bodies full of proteins and sugars. Leaf-cutter ants do drink some of the juice from the leaves they harvest, but they mostly use the leaves they gather as fertilizer for their mushroom farms.

A fungus-growing ant nest hanging from a tree trunk. The pouch-shaped nest is made entirely of mold. Small entryways lead to the inside of the nest. © Jae Choe

A *Gunnera* plant. Common in Central America, it has some of the largest leaves of any plant in the world. After a visit by leaf-cutter ants, the only trace left of the leaves of the plants on the right side of the photo is scraggly leaf veins.

Destruction Above, Construction Below

According to current estimates, there are more than two hundred different species of fungus-growing ants running all different sizes of mushroom farms. Some of them grow their mushrooms on animal excrement or rotten animal carcasses; others gather leaves for their mushroom farms like the leaf-cutter ants. In the La Selva region of Costa Rica, I observed *Apterostigma* fungus-growing ants making mold pouches the size of a baby's fist, which they hung on tree trunks for their nests. They bored tiny holes into the middle of the pouches to haul animal remains through to grow mushrooms on. When I opened one of the pouches I saw a small number of ants living together as a colony. The queen was not much larger than the workers, unlike the leaf-cutter ant colony, but she had slightly a thicker thorax than her workers.

In terms of scale and workmanship, the fungus-growing ants who run the largest and most technically developed mushroom farms are the ants of *Atta* and *Acromyrmex* genera, which are both made up of different species of leaf-cutter ants. There are approximately 40 known ant species in these two genera and they can be found as far north as Texas and Louisiana and as far south as Argentina. Why these flourishing and highly developed ant mushroom farms exist only in the New World and not in other tropical regions would be a fascinating subject of research.

The leaf-cutter ants are much like the Hindu god Shiva, the destroyer and rebuilder of worlds. Once they attack, within a couple of days they can completely strip a tree the size of, say, a shade tree in a park or plaza. Even the destruction that herds of elephants wreak upon the plant life of the African savannahs cannot hold a candle to the damage the leaf-cutter ants can do. Leaf-cutter ants provide one decisive reason why the people of the tropics of Latin America cannot use the same farming techniques used in temperate climates. In a single day, leaf-cutter ants can lay waste to a field that a farmer has tilled faithfully for years. Ecologists estimate that leaf-cutter ants consume as much as 15% of all the leaves in the Latin American tropical rainforests.

However, these ruthless plunderers become the most mild-mannered of farmers underground, dutifully toiling in their highly organized mushroom farms. The Armageddon they wreak above ground is reborn underground as a Garden of Eden. Taking a holistic view of the ecosystem, the leaf-cutter ants' activity is definitely not meaningless destruction. Just as earthworms plow the soil in temperate regions, making it fertile again, leaf-cutter ants also fertilize the soil in the tropics. An average leaf-cutter ant colony plows more than 20 cubic meters of soil; that is 44 tons. Bert Hölldobler and Edward Wilson put this in perspective in their book *Journey to the Ants:* "The construction of one such nest is easily the equivalent, in human terms, of building the Great Wall of China. It requires roughly a billion ant loads to build,

Top, Distribution of leaf-cutter ants in the Americas. Shaded areas show where leaf-cutters can be found over vast portions of North and South America. *Bottom,* Leaf-cutter ants of *Acromyrmex* cutting a lily leaf.

Leaf-cutter ants transporting a leaf.

each weighing four or five times as much as a worker. Each load was hauled straight up from depths in the soil equivalent, again in human terms, to as much as a kilometer."

Mushroom-Farm Assembly Lines

Leaf-cutter ant colonies are highly organized. Divided into castes, the worker ants fall into four groups, distinguished by body size and shape. They range from gardener ants, whose bodies are only two to three millimeters long, to soldier ants that weigh as much as 300 times more and have 6-millimeter-wide heads. As their name implies, soldier ants are dedicated to the defense of the colony. Whenever enemies threaten the colony, the soldier ants mobilize immediately, ruthlessly ripping the invaders to shreds with their mighty jaws. These jaws can tear through human skin or even sturdy leather. One of these soldier ants bit my left pinky when I was digging up a leaf-cutter ant nest in Panama. I could handle the pain, but the bleeding was so profuse that I had to stop work. When I tried to pull the ant off my finger, its neck snapped off.

The head was biting me so hard that it stayed on my finger even after its body had come off.

One caste below the soldier ants are the forager ants, which cut up the leaves and carry them home. The forager ant grabs the edge of the leaf with one of her hind legs and uses her serrated teeth to saw out a circle of leaf to carry home. The forager ant's pace as she transports her leaves is more than just quick. Hölldobler and Wilson, describing this remarkable run of the leaf-cutter workers, estimated that in human terms, the return trip would be comparable to a marathon runner's journey, equivalent to running nearly 15 kilometers at 24 kilometers per hour and for the marathon runner to match the forager ant's trip home, he would have to be holding a 300-kilogram suitcase in his mouth as he ran. The forager ants run their marathon nonstop, day and night, except during occasional rainstorms.

Gardener ants at work in their mushroom farms.

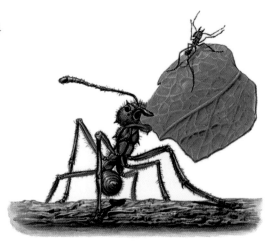

The division of labor by leaf-cutter ants functions much like a factory assembly line.

When the forager ants bring the leaves into the underground farm, smaller worker ants are waiting to cut the leaves up finely with their saw-like jaws, while others take these bits of leaf and chew them into a pulp, much like the pulp we use to make paper, and mix it thoroughly with their enzyme-laden excrement. After thoroughly kneading the leaf paste, they spread it evenly over the dry leaves laid out in the next chamber. Next, even smaller worker ants take little bits torn from the mushrooms grown in the other chambers and plant them in the leaf paste. When the season is right, these newly transplanted mushrooms grow astonishingly quickly—a benefit shared by both ants and humans. Europeans call mushrooms a gift from God because they are well known for how quickly they grow. After a nighttime rain, mushrooms sprout from nowhere, as if planted by nighttime angels.

The gardener ants in the mushroom farms are the smallest workers of the leaf-cutter ant colony. The gardeners' chores include keeping the area around the farm clean, weeding, and harvesting the mushroom crops. From cutting up the leaves and carrying them underground to raising and harvesting the mushrooms to storing the crops, the worker ants' workplace is reminiscent of the assembly lines of modern-day human factories. The leaf-cutter ants' system, with each ant engrossed in a specific job based on her body size, is the epitome of division of labor.

Mushroom Spores as Dowry

When the queen founds her new colony, she must begin without any help. But where can she get mushroom spores for her new farm? In many human cultures, women receive money or other valuables from their families to take with them as dowry when they leave home to marry. Similarly, when the leaf-cutter ant queen leaves home for her nuptial flight, she carries a cluster of mushroom spores in a small pocket inside her mouth. After the queen mates with the males, she picks a good spot to dig a nest for her new colony. Once inside, she immediately spits out the mushroom spores and starts tilling the soil for a new mushroom farm. When her worker ants hatch, they bring in leaves from outside, and the small garden the queen has prepared quickly expands into a massive farm.

Entomologists of social insects have discovered that through this matrilineal tradition of passing mushroom spores, leaf-cutter ants have passed the same species of mushroom, via mother to daughter, to every leaf-cutter ant colony. All leaf-cutter ants grow the same kind of mushroom in the southern United States and throughout Central and South America. It has also been determined, through DNA analysis, that this relationship between leaf-cutter ants and the mushroom species is nearly 25 million years old. This is similar to the way yogurt-producing bacteria have been passed from one generation to the next by humans. Perhaps the early leaf-cutter ants tried growing other mushrooms that grew in their habitats, but after finding the ideal breed of mushroom they developed a tradition of only growing that and nothing else. It must be one of the world's longest running traditions.

From a Single Room to a Metropolis

Leaf-cutter ants form colonies on a staggering scale, ranging from 5 to 8 million ants per colony. One colony dug up in the rainforests of Brazil contained more than thousand chambers, ranging from the size of a fist to the size of a soccer ball, and the ants were growing mushrooms in about 400 of these chambers. A single queen can lay up to 150 million eggs in her lifetime. Some of these eggs will hatch the queens and males that will found the next generation of colonies, but most of them will hatch worker ants.

In order to lay this many eggs, the queen must mate with multiple males during her nuptial flight. She will store at least 200 million sperm. The queen stores the sperm away in pouches in her body called spermatheca and she must use these sperm for the rest of her life. How long can the leaf-cutter ant queen live? We do not know exactly, at least, not in their natural environment, but a queen ant that Hölldobler and Wilson raised in their laboratory at Harvard University lived for at least 14 years. How the queen keeps the sperm healthy for so long, at body temperature, would be an interesting physiological study. The human technology of storing sperm in liquid nitrogen for future artificial insemination cannot compare to what the queen ant can do.

After her nuptial flight, the queen ant digs a tunnel 12–15 millimeters in diameter, as far as 30 centimeters deep, to a spot where she makes a small space for herself that is about 6 centimeters in diameter—a small room that may look insignificant, but in the days to come it develops into a metropolis with millions of citizens. This metropolis contains not only rooms to cultivate mushrooms but also central heating and cooling systems interlaced with a complicated series of air ducts for ventilation and large rooms deep underground where decaying plant matter and other debris are stored. The hot air that rises from the rotting garbage flows out of the colony via air ducts leading to the surface, and at the same time oxygen-rich cool air flows in through other passageways, providing ventilation and maintaining the internal temperature. With their factories based on a division of labor as specialized as that of an assembly line, as well as the wide variety of infrastructure in their well-managed cities, leaf-cutter ant societies are astonishingly evolved. Only human civilization can compare in terms of systematic productivity.

4

Ant Ranchers

*masters of dairy farming second
only to mankind*

No one knows exactly when humans first started raising live-stock—cows, pigs, and other now-domesticated animals. Human beings breed animals in order either to directly consume the bodily tissues, as in the case of beef, pork, and chicken, or to snatch the eggs, milk, and other materials that animals produce for their own reproduction. In order to satisfy our enormous appetites, human beings have vast pastures, cattle ranches, and poultry and fish farms that we use to raise and breed all kinds of creatures. In order to overcome the scarcity of food in Latin America, people have even resorted to farming iguanas and large rodents such as agoutis and capybaras.

Ants and Aphids

After human beings, ants may well raise more livestock than any other creature in the animal kingdom. Ants and aphids are frequently cited as a good example of a symbiotic relationship—different species living together and providing benefits to each other. Ants protect aphids from their natural enemies such as ladybugs and green lacewings and the aphids in return provide their ant protectors with some of the nutrients that they suck out of plants in the form of the honeydew.

If we observe the behavior of ants and aphids carefully, we see that the relationship between ants and aphids is not always so mutually beneficial. British researchers have observed that aphids spend 14% of their day under the protection of the ants, but they produce 84% of their honeydew during this time. In other words, the honeydew that aphids produce is almost entirely for ant consumption.

One aphid does not provide enough honeydew to feed one ant. Looking at the colony as a whole, however, we see that the honeydew from the aphids that the worker ants raise individually as well as those they herd together provides as much as 75% of the nutritional needs of the colonies that raise aphids. These ants are indeed masters of dairy farming.

When we analyze the contents of aphid honeydew, we can see how much they appreciate the ants' protection. Aphids do not just suck the juice out of plants and give it directly to ants. The honeydew that aphids provide to ants contains water, carbohydrates, many different amino acids, and other nutrients in just the right amounts, making it a balanced diet for ants.

Ants do more than simply protect aphids. Just as shepherds herd their flocks to places with plenty of grass, ants have been known to herd their aphids from leaf to leaf. They find the best spots on the plant for the aphids to suck the juices and transport the aphids there.

Ant milking an aphid for honeydew.

Some ants do not just let their aphids roam the pastures but actually build stables to raise them in. These ants build dugouts at the roots or stems of plants to house their aphids. While I was studying for my master's degree at Penn State University I would sometimes drive in the entomology department's old truck with Peter Adler, now a professor at Clemson University, to do field observation in the neighboring woods. At the base of a meter-high poplar tree we discovered a mud dugout surrounded by a mud wall inside which ants were tending a herd of aphids. The next four days we returned to observe the ants and aphids. Then suddenly both ants and aphids disappeared without a trace. All the dugouts had been removed in a single morning. This happened

A *Paraponera* worker ant returning to the nest, mouth filled with honeydew.

Citronella ants (*Acanthomyops*) raising scale insects in stables inside their nest. Squeeze these ants gently with your fingers and a lemony scent is given off by their abdomens.

during the last week of June, and even more amazingly the ants built similar dugouts there at the almost the exact same time the next year. The year after that I transferred to Harvard so unfortunately was not able to continue our research on that project. To this day I sometimes wish we could have found out whether or not this phenomenon occurred again.

Many Different Species of Livestock

Humans milk not only cows but also goats, sheep, and camels; and ants, too, raise many different kinds of livestock. Mostly they raise aphids, but they also tend other kinds of insects, such as treehoppers, leafhoppers, scale insects, and other insects belonging to the same order of Homoptera that aphids belong to, as well as caterpillars from the Lycaenidae family of butterflies. Ants let most of their livestock roam free on tree trunks and leaves, but scale insects they keep inside their nests.

For reasons that are not clear, the further into the tropics one goes the harder it is to find aphids. Likewise, it becomes harder to find ants that raise aphids, so tropical ants raise a variety of different kinds of livestock. These ants mostly raise treehoppers, whereas ants from temperate regions mostly raise aphids.

One day, deep in the tropical rainforests of Panama, I discovered an Aztec worker ant that appeared to be sucking something out of a tree. Looking closely, I saw, sticking out of the tree, a small, white pointy tube that the ant was poking with its antennae. Each time the ant poked, a small droplet came out of the tube. I cut off the bark surrounding the tube and found a small, pale-colored scale insect inside. The scale insect was obviously being protected by the ant and was providing honeydew to the ant. Another way of looking at it was that the ants had confined the scale insect to a tiny stable in order to milk it. The ants seem to be as cruel as human beings, shutting their livestock away in tiny spaces with no concern for their animals' welfare, only for what they could harvest.

Bees and wasps also engage in dairy farming, though less commonly than ants. On the Pacific coast of Costa Rica, along the Panamanian border, there is a virgin tropical rainforest called Corcovado. This forest cannot be reached overland, only via light aircraft. In the summer of 1984, I, with Deborah Letourneau from the University of California, Santa Cruz, observed ants and wasps competitively protecting and milking leafhoppers in Corcovado. What we found especially interesting was that the leafhoppers preferred the wasps to ants. When the wasps went away, to take the harvested honeydew back to their nests, ants appeared and started milking the leafhoppers. However, the leafhoppers gave

Opposite, Tropical ants raising various kinds of leafhoppers. The middle photo captures the moment when an ant is taking a drop of honeydew from the rear of a treehopper.

much less honeydew to the ants than they gave to the wasps, and they gave it far less often. They appeared to prefer the wasps because the wasps were bigger and could protect the leafhoppers better than the ants could. The livestock appeared to have picked the farmers they liked best.

Coevolution with Ants

Most of the insects that ants raise as livestock are smaller, herbivorous insects with no defense mechanisms of their own. Most ants are predators, and many ants routinely capture and consume a wide variety of herbivorous insects. How and when some of these insects came to be companions to ants instead of becoming the ants' prey, helping the ants to thrive, is unclear, but it is interesting to investigate such mysteries. Instead of preying on them, these ants protect their livestock insects from other predators in return for the nutrients they provide.

Aphids suck out a large quantity of plant juice to obtain amino acids. When they suck out too much sap they have to release some of it. During this process, they excrete not only water but also some of the sugar from the juice comes out with it. As a result, the area around them becomes sticky and starts to smell, which attracts predators, disease-carrying insects, and even mold. The ants that protect aphids help keep their bodies and surroundings clean.

Why, then, don't all aphids have symbiotic relationships with ants? Interestingly, the bodies of aphids that live with ants have evolved differently from the bodies of aphids that do not. Aphids that are protected by ants have long legs, and their cornicles, or honeydew-extruding tubes, are also long. Aphids that are not protected by ants have shorter legs and cornicles. In addition to having made such physical adaptations, aphids that live with ants also

Opposite, Aztec ant milking a scale insect shut away in a tiny stable (*top*). The cover of the stable was removed for photographing.
Wasp getting honeydew from a leafhopper as a reward for its protection (*bottom*). © Jae Choe

Treehopper mothers often leave their offspring with a wasp nanny so they can have more babies elsewhere. © Jae Choe

behave differently. They do not constantly release their honeydew at a fixed rate; they produce it, in concentrated bursts, only while the ants are protecting them.

Some kinds of treehopper even take advantage of the ant child-care service by completely relying on ants to look after their off-spring. Some mother treehoppers entrust all of their young to trustworthy ants so they can take off to lay more eggs. A mother treehopper that leaves her babies with ant nannies can have a lot more offspring in her lifetime than a treehopper that takes care of her own babies. There is a big evolutionary advantage to this practice. We might even say that these livestock seem to be using their owners.

5

The World's First Bodyguards

It is widely understood even by people with scant knowledge of natural science that plants and animals have symbiotic relationships; for example, that honeybees and butterflies flit from flower to flower transferring pollen in return for nectar. Unlike animals, the plants do not have the luxury of being able to wander around looking for a mate: they have to rely on flying animals such as bees, butterflies, birds, and bats to help them. This is an indirect sexual relationship. Since they cannot negotiate actively with their lovers, they rely entirely on selections made by their visitors in order to reproduce. To tempt partners in this exchange, they open wide their sex organs, beautiful flowers for the whole world to see, and offer sweet nectar from deep inside to reward the visitors' efforts.

In terms of structure and function, the flower is the product of coevolution. Flowers have evolved over time to help make it easier for bees and other animals to pollinate them. Ants, though, who in the evolutionary scheme are closely related to bees, play nowhere near as important a role in pollination as bees do. According to current research, they serve as pollinators for only a few kinds of plant. The flightless worker ants that make up the majority of ant colonies cannot spread pollen as far as bees, and plants need to spread their pollen as far as possible. Ants may be cheaper—plants may not have to give as much of a reward to ants as they might give to bees—but plant priorities rate ants as not the best way to get pollinated.

Plants Using Ants as Guards

A surprising number of plants have nectaries on their leaves or stems as well as inside their flowers. These nectaries are called extrafloral nectaries, and unlike the floral nectaries, designed to attract bees and similar pollinators, extrafloral nectaries are almost always provided for the benefit of ants. These nectaries provide ants with nectar that is rich in concentrated sugars. In return, ants protect these plants from herbivorous animals.

Interestingly, while the nectar that comes from floral nectaries is rich in protein as well as sugars, the nectar that comes from extrafloral nectaries contains almost no protein. This means that the ants that feed on this nectar must prey on herbivores for their protein. The nectar that comes from the floral nectaries of the balsa and that is taken by pollinating bats consists of 11% sugars and amino acids. The nectar from the extrafloral nectaries on the same trees, however, contains almost no amino acids.

These plants have struck a reasonable bargain with the ants that protect them. The amount of energy they use to maintain one of these extrafloral nectaries is equivalent to what it would take to grow approximately 13 square centimeters of leaf area. This is no more than 1% of their leaf area. Herbivores could easily consume much more. If the ants that use a plant's extrafloral nectaries can reduce the amount of leaf area consumed by herbivores by more than 1%, the benefits to the plant outweigh the costs of the arrangement.

Nectar does not flow constantly from these extrafloral nectaries. Each plant secretes its nectar at different times. Some plants secrete nectar only when they grow the buds that herbivores like to eat—that is to say, when they need to summon ants. Other plants produce nectar when their seeds are mature so that ants can chase away seed-nibbling insects.

Some Plants Provide Ants with Shelter

The more time that herbivore-hunting ants spend running around on their leaves and stems, the better the protection that the host plant receives. Some tropical plants even go so far as to provide

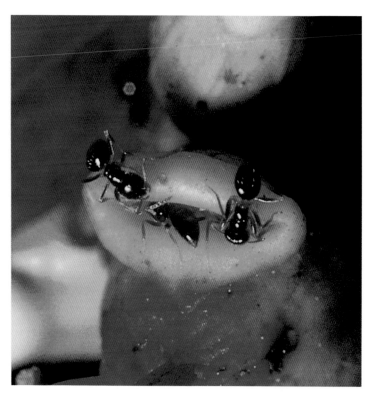

Costa Rican *Brachymyrmex* ants sucking nectar out of a poinsettia's extrafloral nectary.

not only food for their ant protectors but also a place where the ants can live. The bull's horn acacia of South America is a typical example. The bull's horn acacia has thorns like most species of acacia, but the thorns are unusually large and are shaped like the horns of a bull. *Pseudomyrmex* ants make their home inside these thorns.

Like other plants, the bull's horn acacias provide nectar and shelter for ants, but they also produce packets rich in protein and fat—called Beltian bodies—on the tips of their leaves. These Beltian bodies, consumed by ants, are so nutritious that the ants do

A *Pseudomyrmex* worker ant making its home inside the thorn of a bull's horn acacia (*top*); clinging to the end of a bull's horn acacia to harvest a Beltian body (*middle*); and harvesting nectar from the bull's horn acacia's honeypot-shaped extrafloral nectary (*bottom*).

not need to eat herbivores to meet their needs. However, the *Pseudomyrmex* ants that live on these trees are so ferocious that not only herbivorous insects but even deer, cows, horses, and other large herbivores do not dare go near the bull's horn acacia trees.

Pseudomyrmex ants not only protect the trees but they also keep the area around the acacias clear of other plants, allowing the acacias to grow more quickly. The *Pseudomyrmex* ants residing in younger bull's horn acacia trees are dedicated to keeping the area around their acacia trees clear of any obstructions to the sunlight they need to grow. The worker ants are so devoted to clearing out other plants as quickly as they sprout that it is hard to find even a single blade of grass surrounding these trees. They are constantly weeding for their acacias. Thanks to the *Pseudomyrmex* ants, bull's horn acacias have almost no competition for the sunlight they need to flourish.

Professor Dan Janzen of the University of Pennsylvania, who is often considered to be the godfather of tropical biology, proved through his field experiments what indispensable companions the *Pseudomyrmex* ants are to bull's horn acacias. He sprayed insecticide on bull's horn acacias that were under the protection of *Pseudomyrmex* ants and observed changes to their surroundings. Shortly after he stripped the acacias of the ants' protection, other plants began to sprout up around the acacias, competing with the acacias. Not only that, but a variety of herbivorous insects began attacking the trees. *Pseudomyrmex* ants and bull's horn acacia trees evolved into their mutualistic relationship long ago and have continued to maintain it to this day. Occasionally birds and wasps build nests in bull's horn acacia trees, which is further testament to how effective the *Pseudomyrmex* ants' protection is.

More than two decades ago, Dan Perlman and I conducted research in the tropical rainforests of Costa Rica on the mutualistic relationship between Aztec ants and *Cecropia* trees, also known as trumpet trees. Like bamboo trees, trumpet trees are hollow, and Aztec ants make homes inside these hollow trunks. Like the bull's horn acacia, trumpet trees provide ants with food and shelter. They also grow glycogen-rich nutritional packets, called Mül-

Top, Weeding by *Pseudomyrmex* ants helps bull's horn acacia trees to grow without competition. *Bottom,* Bull's horn acacia trees are sometimes chosen as nesting sites by birds—a choice that testifies to the effectiveness of the ants' protection.

Opposite, When bull's horn acacia trees do not have *Pseudomyrmex* ants to protect them, competing plants grow around the acacias.

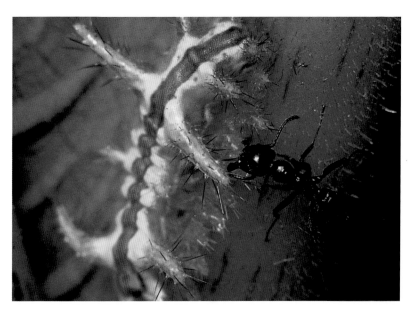

An Aztec ant attacking an herbivorous insect it caught nibbling on a trumpet tree.

lerian bodies, at the base of their leaf petioles. The ants reward the trumpet trees for providing them with food and shelter by protecting them from herbivores and competitor plants.

Ants Even Spread Seeds

Since plants are fixed in a single location, moving their seeds to places with little competition can be a problem for them as serious as that of spreading their pollen. Dandelions solve this problem by letting the wind carry their seeds; there are even plants that shoot their seeds out of their ovaries by bursting them like balloons. Many plants hide their seeds inside pleasant tasting fruits so animals will eat the fruit and later, in their excrement, deposit the seeds elsewhere. The fruits we eat all come from plants that use this method to spread their seeds.

Top, There are plants that grow only in ant gardens. *Bottom,* Trumpet trees provide these Müllerian bodies to Aztec ants as food.

Some plants, however, rely entirely on ants to spread their seeds. Celandines, a kind of poppy that grows wild in many parts of Korea and throughout East Asia, are a good example of this. These plants' seeds have a special attachment called an elaiosome that is rich in fat. Ants take these seeds back to the nest to eat the elaiosomes and then toss the rest of the seeds intact into the colony's garbage dump. Ant garbage dumps usually contain a lot of other food remnants, so when the seeds reach germination stage the nutrients in the rotting food waste let them sprout quickly. These plants have also coevolved with ants. In their mutualistic relationship, these plants and ants have grown so close that the plants' seeds can germinate only in ant garbage dumps or ant gardens. The celandine has become a flower that only ants can grow.

6

The Charge of the Ant Brigade

. . . .

the terrifying march of the army ants

The razor-sharp, sickle-shaped jaws of the ants snap merci-lessly as the giant insects swarm the village. The village be-comes bedlam, the villagers climbing over each other in their desperation to escape. Those who cannot outrun the ants are cut to pieces by the massive jaws. A scene from a Hollywood science-fiction movie? It certainly could be.

My life in the United States always revolved around a natural history museum on university campuses. I worked in the Frost Entomological Museum while I was studying for my master's degree at Pennsylvania State University, and later it was the Museum of Comparative Zoology at Harvard University, where I was doing my PhD. At the University of Michigan I actually had my office in the Museum of Zool-ogy. At all three museums I sometimes acted as a tour guide for kindergarten and elementary schoolchildren. Before I showed them our exhibit of the Hercules beetle, the largest species of beetle in the world, I would ask what the largest insect they had ever seen was. They would almost always mention giant, man-eating ants—referring of course to the notorious army ants. As terrifying as army ants are, though, they generally do not harm people, except for the occasional newborn baby left in the wrong place at the wrong time. For smaller animals in the grasslands of hot climates, however, army ants often are a real, mortal danger.

Army Ants Take No Prisoners

Army ants can be found worldwide in the tropical and tem-perate regions of North and South America, Africa, Austra-lia, and Asia. In the Americas alone, there are more than 140 known species of army ants. The best-known species of the army ants is *Eciton burchelli,* which inhabits the tropical rainforests of Latin America. When conducting various kinds of field research on Barro Colorado Island in Panama, where I spent many years, I observed the systematic raids of the army ants many times. When army ants came, particularly *Eciton* army ants, I could hear them before I could see them. In the morning around 10 o'clock, if I listened carefully I could hear birds making a commotion deep within the forest. It was the sound of antbirds getting excited. These birds follow army ant swarms so they can snap up panicking ground insects as they try to escape from the marauding army ants.

Army ant raids are a much-awaited feast for not only the antbirds who follow the army ant trails but also for beautifully winged

Eciton army ants attacking a large scorpion.

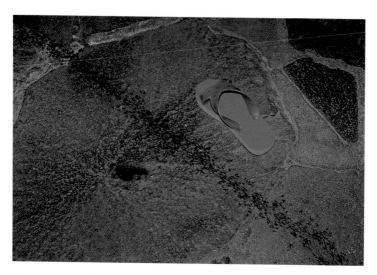

An army ant trail passing my colleague's sandal, which fortunately he wasn't wearing.

army ant butterflies that feed on antbird droppings. Parasitic flies add to the din, boisterously buzzing around the ant trails, swooping down on the army ants to lay their eggs behind the ants' necks. For a neutral observer, an army ant raid is quite a show; for the animals affected by the raids, these are times for savage clashes of life and death, joy and sorrow.

Army ants typically feed on cockroaches, beetles, grasshoppers, spiders, and even scorpions. Cockroaches form the mainstay of the army ants' diet. While human households fret about cockroach infestations, it would mean a major food shortage for many tropical ecosystems if cockroaches ever disappeared. During a raid, groups of ants will stop for a few hours at a time, clustered around life-sustaining cockroaches and stick insects. After gnawing off pieces of this plunder and transporting them to the rear of the swarm, the army ants continue their raid. Once the army ants have finished raiding an area, it is hard to find any animal life in the desolate ruins they leave behind.

A Life of Constant Wandering

Because of their destructive habit of turning the places they visit into wastelands, army ants cannot stay in the same place for long. Unlike other ant species that pick sunny places to build highly organized underground cities, army ants are nomadic, constantly wandering. After they plunder an area, they build temporary nests, called bivouacs, to camp in for the night. They do not actually pitch tents, but worker ants hold on to each others' feet to form a curtain of bodies inside which their fellow ants can rest. The queen and larvae make camp in the center of the curtain. A bivouac curtain consists of at least 500,000 ants and can weigh as much as a kilogram. When daylight breaks, the ants wake up and hit the trail again.

According to current research, ants on a raid do not seem to follow a leader. Once they leave on a raid they press forward at speeds of approximately 20 meters per hour. Researchers have also discovered that during raids each species of *Eciton* army ant has its own tactics, differing slightly from other species. Ants of *Eciton burchelli* travel together in several, though just a few, columns and then suddenly fan out to plunder a larger area. Other species gradually branch out from a single column, occasionally forming smaller fans at the end of each branch. Other species' formations lie somewhere between these two extremes.

Organized Societies with Divisions of Labor

Nearly all ant species are efficient predators, but there is a limit to how much prey a single ant can catch. One ant can catch and carry only small prey, and even if it catches something larger, it needs to call in other ants to help carry it. Army ants, however, have no such problem. They attack in such large numbers that they can easily finish off larger prey. With a few exceptions, like animals that can outrun the marauding army ants or some ticks and stick insects that can chemically camouflage their bodies, almost every animal that falls within the path of the army ants will be eaten.

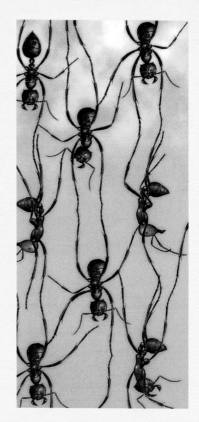

Right, Bivouacking army ants, their bodies forming a curtain. They do this by holding on to each others' feet. Below, Patterns of army ant trails. They travel together in single file until they reach an area with a lot of food, at which point they fan out.

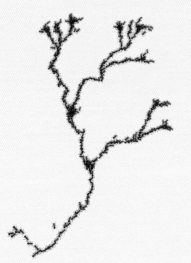

Top, A major, the largest of the army ants. The majors' job is to protect the colony from enemies. This one, with large, razor-sharp, sickle-shaped jaws, is poised to attack. *Bottom,* A submajor worker ant, smaller than the majors, guards the even smaller media workers.

Opposite, A submajor worker ant watching over media worker ants carrying pupae. The jaws of each type of worker ant are shaped differently.

Watch an *Eciton* army ant trail closely and you will notice a few larger, lightly colored worker ants running back and forth among smaller, darker-colored workers that mostly march straight ahead. These larger ants are called majors; they are soldier ants, with frightening jaws shaped like ice tongs. They do not help the rest of the colony with catching prey or transporting it; their job is to protect the colony from enemies. Their sharp jaws, curved like sickles, are not suitable for carrying food.

There are four classes of worker ants. A colony has only a few majors and submajors, the largest being the majors. Then come the submajors, which are somewhat smaller and have lighter-colored

heads and thoraxes. Submajors carry food to the rear of the raiding party. Their jaws are designed for carrying. Up to two-thirds of the colony's worker ants are medium-sized, called media ants. The media serve as general laborers, responsible for most of the colony's work, including transporting the colony's larvae and pupae.

Ants with No Nuptial Flight

The smallest workers of the army ants are the minors. These serve primarily as caretakers for the queen, who is so fat she can barely move. Army ant queens do not make nuptial flights. Instead, in breeding season the males leave the nests they were born in to find another nest's queen to mate with. Male army ants are approximately the same size as the queens, and unlike the males of other ant species army ant males break off their wings when they find a queen to be their mate.

Army ant queens have pheromones that are unique to each queen, and the queen uses these pheromones to control worker ants. Army ant males have evolved similarities with the queens: their bodies are full of glands that secrete queen pheromones. It is believed that they have developed this ability to copy pheromones as a survival strategy, and that the secretions enable them to sneak past the vicious worker ants of other army ant colonies.

Eciton army ants alternate cyclically between a nomadic phase and a stationary phase approximately every 35 days. After about 15 days of nomadic life they bivouac, lay eggs, and raise the next generation of worker ants. During the 20 days of the stationary period, the queen lays and hatches between 100,000 and 300,000 eggs. If a colony becomes too bloated after several repetitions of this cycle, it divides into two colonies. After the males have sneaked in to spread their sperm, the young queens compete to see which queen gets to lead the new colony. The two strongest queens divide the colony in half and march in different directions, and queens who have been left behind are detained by some of the workers. The queen who fell from power and the queens who lost the competition are then ruthlessly left to die—a fate not so different from what happens in politics in human societies.

PART II

The Culture of Ant Society

7

Talking with the Ants

*the clever designs of ant
communication*

In the science-fiction novel *Empire of the Ants* by French author Bernard Werber, humans learn the language of the ants and open lines of communication with them. Modern biology is developing at breakneck speed: could the day actually come when communication between ants and humans is possible? Should this day come, whose language would we speak?

Animals communicate through their senses of sight, hearing, smell, and touch, and they can perceive information through vibrations, using their organs of hearing and touch. We humans are particularly dependent on our senses of sight and hearing. For example, we usually recognize each other visually, rather than through smell or touch, and we also use visual clues to understand each other, such as facial expressions, physical gestures, and written words. Of course, the bulk of our communication comes in the form of a wide variety of spoken languages.

Humans rely on their eyes, ears, and mouths to communicate. For the rest of the animal kingdom, however, the sense of smell is usually the most important medium of communication. We can see an everyday example of this in the neighborhood dogs, sniffing each other when they meet on the street to determine each other's identity and condition. Human beings, by introducing ourselves with words and swapping business cards, are the unusual ones. That said, it does not mean that we lack the sense of smell all together. We may not realize it, but at some level we also rely on our sense of smell to communicate—a fact that keeps perfume companies in business.

The Dance Language of the Honeybees

Unlike the environment of many science-fiction novels, the world portrayed in *Empire of the Ants* is neither impossible nor absurd. Humans have, in fact, already established a level of communication with honeybees, which have at least as high a level of social organization as ants. Honeybees are well known for their ability to communicate with each other through dance. In 1973, Karl von Frisch, along with Niko Tinbergen and Konrad Lorenz, was awarded the Nobel Prize in physiology or medicine for his pioneering work in the field of modern ethology—that is, the study of animal behavior—and especially for deciphering the dance language of bees.

When a scout bee discovers flowers filled with nectar, she returns to the hive and performs a "waggle dance" to inform her nestmates of the distance and direction of the new source of nectar. The duration of the waggle dance shows how far away the source of food is and the angle of the dance on the vertical surface of the nest's comb indicates the direction in which the food can be found, relative to the sun. The information the bees convey through this dance is so precise that even humans can accurately determine the location of the bees' sources of nectar by watching them dance. We have at least learned how to read the bees' language.

More than that, however, a group of German scientists from Max Planck Institute managed to send messages to bees. They built tiny robots to perform a waggle dance for bees and successfully sent the bees to a predetermined location. Communication between bees and humans has thus become possible, even if the communication, both coming and going, is driven by only one side, the human. Were bees to realize that we can understand their language, perhaps they would approach us to say more.

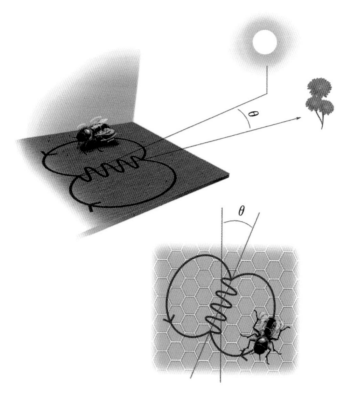

Top, The scout bees, finding a nectar source, tell other bees where to find it by using their angle of dance in relation to the direction of the sun. *Bottom*, Diagram of a bee's "waggle dance." The duration of the dance indicates distance to food source.

Ants Speak through Scent

Like most insect languages, ant languages are chemical-based. The next time you see an ant carrying food back to the nest, try watching it closely from the side. You will be able to notice that the ant drags the tip of its abdomen against the ground as it moves along. It is leaving a chemical trail, also known as an odor trail, which leads an ant from a food source back to its nest.

If an ant encounters other ants from its colony as it leaves its chemical trail on the way back to the nest, it lets them sample some of its cargo. After tasting the food, the other worker ants will follow the chemical trail to the source of the food. An old Korean proverb says, "Even an ant crawling by makes its mark," meaning that even the most insignificant thing can have a lasting effect. It seems ancient Koreans realized that ants leave odor trails.

The chemical that ants deposit on the trail is a pheromone, one of a wide variety that ants' bodies can produce. Every ant's body is a walking industrial complex, equipped with chemical factories from head to the tip of the abdomen. Determining which gland each ant chemical came from was a relatively simple matter. After dissecting and separating each individual gland, scientists drew trails with each gland in different directions away from the opening of an ant nest. From there, it was easy to see which gland produced the pheromone that tells worker ants to follow it to find food. It was not always so simple, however; ants often use more than one gland at a time to produce chemical cocktails of pheromones.

Ant chemical language is much more economical than our verbal language. The pheromones that the leaf-cutter ants secrete from their stingers are chemically complex, and we have created a suitably complex word for the name of the chemicals the ant pheromones are composed of: methyl-4-methyl-pyrrole-2-carboxylate. However, ants are so sensitive to this chemical that if even one milligram of it were stretched into an odor trail that went around the world three times, ants would still be able to detect and follow it. It is the instability as well as the potency of the pheromones in ant chemical trails that makes the trails such an efficient means of communication. If the trails did not dissipate shortly after all of a food source had been collected, worker ants would keep following the trails in vain, erroneously believing that food was waiting for them. This means that worker ants following a chemical trail to take food back to the nest must leave more pheromones on top of the parts of the trail that are already fading. Only when ants start returning to the nest empty-jawed do they stop depositing pheromones on the trail, leaving it to dissipate naturally.

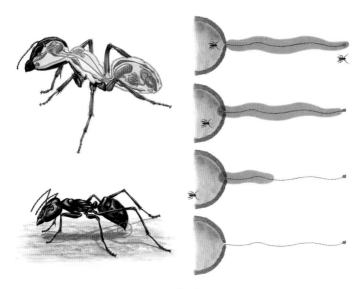

Left, The ants' secretory glands (*top*). Ants leave their chemical trails by dragging the tips of their abdomens along the ground (red circle, *bottom*). The ant secretes its pheromones from here. *Right,* Once the food source is exhausted, the trail starts to dissipate.

Chemical Alarms

When invaders enter their nests, ants sound a chemical alarm. I was able to observe this firsthand during my research at the Smithsonian Tropical Research Institute on Barro Colorado Island. Certain species of the Aztec ants that are common in many parts of Latin America make their nests along large tree branches that are about two meters (six feet) long. These ants are so ferocious that if you hesitate near the ants' trees for too long, the Aztec worker ants will swarm and attack you.

As soon as they discover an insect interloper they secrete alarm pheromones, and swarms of workers from the colony rush to the scene to surround the intruder. Once the workers grab all three pairs of legs and the antennae, the intruder's fate is sealed. The workers pull in all eight directions, tearing the enemy apart like a medieval prisoner being drawn and quartered.

Aztec ant menacingly rocking its head back and forth at the camera—at least, photographers sometimes think the menace is aimed at them. © Jae Choe

We know a great deal about the chemical language of the weaver ants of Africa, Southeast Asia, and Australia, thanks to the extensive research conducted by Bert Hölldobler and E. O. Wilson. The weaver ants work together to make their nests by pulling leaves of the trees close and sewing them together, using the silky thread that their larvae spin. In order to accomplish this complex task, even mobilizing their young, the weaver ants' highly developed social organization and chemical language are absolutely necessary.

Marking territory and informing the rest of the colony about the presence of intruders or the location of food are just a few of the simpler messages they express in their chemical language. Social insects such as weaver ants have developed ways to put together "words" to create a variety of "phrases," which makes the high level of organization of their society possible. We like to believe that humans have a monopoly on language, but the language that the weaver ants use to communicate with each other has formed the basis of their highly complex society.

Stomping Their Feet to Communicate

Ant communication is mostly based on chemical languages that use the ants' sense of smell, but their senses of hearing and touch also play important roles. Ant biologists' research over the past 20 years has revealed that many species of ant use sound to communicate. Insects like crickets, grasshoppers, and cicadas are well known for the sounds they make to communicate, but the sounds the ants make are inaudible to the human ear. If the ant in the "The Grasshopper and the Ant" were to sing, we would not be able to hear it.

The more primitive ant species tend to rely heavily on touch for communication. When these ants want to lead other ants from the colony to where they have found food, they must do so one ant at

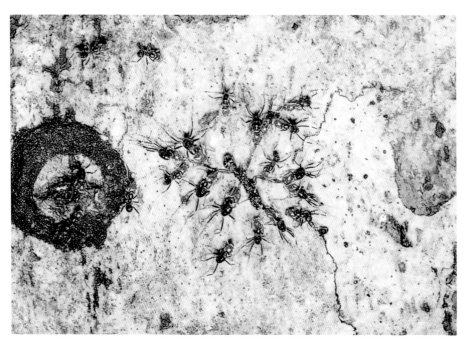

Aztec ants tearing an invader limb from limb. © Jae Choe

a time. To do this, a worker ant first finds another worker ant and taps on it a few times with its antennae, then turns around and walks toward the food. The second worker ant follows closely, almost touching the lead ant. Sometimes, the second worker ant loses track of the lead ant, and if the leader can no longer feel the follower ant's antennae on its rear end, it stops, turns around, and finds the follower ant. If the follower stops following, the lead ant might even bite her and drag her along. An interesting subject for further research would be how ants evolved from this inefficient form—ant-to-ant communication, called tandem running—to quickly mobilizing large numbers of worker ants through mass communication, leaving chemical trails.

8

The Ants Come Home

sense of direction and
biological clocks

The story could happen only in the movies. *Lassie Come Home* is a classic, and probably everyone has heard about Lassie, the dog that overcame all kinds of adversity on her long journey as she finds her way home. But dogs actually mark their paths with their urine as they go, and this works only to guide them for short distances. Lassie had to travel much too far to have been able to leave a urine trail as she went.

Fiction or not, an elderly couple was so touched by *Lassie Come Home* that they donated a considerable sum of money to Harvard's biology department in order to research the question of how a dog could have traveled such a long distance to find her way home. Obviously, no one used this money to research Lassie's behavior. Nevertheless, the endowment did subsidize quite a few students like me who were studying various forms of animal behavior.

Using the Sun and the Stars for Guidance

Pigeons' amazing homing abilities work differently than dogs. If you catch a pigeon near its nest, put it in a box and set it free in a place it has never been to before, it will still almost always find its way home. Extensive research by behavioral scientists has revealed that pigeons use many different kinds of sensory input to guide them. If the distance is not too great, they can use their sense of smell to find olfactory clues in the air or they can orient themselves visually using landmarks on the ground to find their way home. Over longer distances, however, they must rely on the sun or the stars for guidance.

In order to stay more or less on their odor trails, ants glean through their antennae.

Some people get lost in the woods even though they have a compass. A compass may tell you which direction is north, south, east, or west, but it cannot tell you which way you want to go. In order for any guidance system to be useful, you first have to know which way your destination is in relation to your current position. Pigeons are able to use the sun and the stars to figure out which direction they are going, but how do they know what direction to go to find their home? Pigeons use the Earth's magnetic field to determine what direction their destination is from their current position. Scientists proved this by attaching small magnets to the tops of pigeons' heads to disrupt their ability to detect the Earth's magnetic field. The pigeons in these experiments could not find their way home.

How do ants navigate? With the exception of some of the more primitive species of ants, they follow the chemical trails that they or other ants from their colony have left behind, leading from the nest to places that have food, and back again. These pheromones tell the ants where the trail is, but they cannot indicate which direction to go on the trail: which way is in and which way is out. It could be that the density of pheromones at different points along the trail indicates direction, but current research suggests that this is not the case.

This is because there is probably a limit to how well ants' antennae can determine the "thickness" of the chemicals on the trails they are following. Ants balance the amount of olfactory stimulus they receive from each of their antennae so that they can follow their chemical trails. If they stray from the center of the trail, the stimulus they receive through the antenna on one side will not be as strong as the stimulus they receive from the antenna on the other side. As the stimulus becomes weaker on one side, the ant weaves, left, right, constantly changing course as it follows the trail.

Human nostrils are too closely attached to each other for us to compare the olfactory stimuli that come in simultaneously through both sides. We can, however, hear the difference in the volume of the sounds we hear through each ear, and this allows us to turn in the direction a sound is coming from. In a similar way, ants stagger along their trails as if they were drunk. They are constantly comparing and evaluating the stimuli they receive through each of their antennae. This goes on from one moment to the next: they do not seem able to remember the stimulus from even a moment ago or be able to judge whether a stimulus they are currently receiving is stronger or weaker than the stimulus they received earlier.

If you pick up an ant that is on its way back to the nest in the dark and turn it around to face the opposite direction, it will usually keep going in that direction, away from the nest. A Swiss entomologist named Felix Santschi discovered in Tunisia nearly a century ago that ants actually use the sun to determine which direction to go. Hypothesizing that ants might use the sun as a compass to help them find their way back to the nest after foraging for food, Santschi devised a simple experiment. He set up a board to block the sunlight next to the trail of a group of ants that were taking food back to their nest. Unable to see the sun, they stopped going forward and started running around in confusion. As soon as he removed the board the ants began to head back to the nest again.

To test his hypothesis, Santschi moved to the next stage of his experiment. He placed a mirror across from the boards he used to confuse ants that were on their way back to the nest. The reflection of the sunlight in the mirror made the ants believe the sun was in the opposite direction and they turned around and started going away from their nest. Once they left the area with the artificial sun between the board and the mirror, they turned and headed back toward the nest until they were between the board and the mirror again. At that point they yet again turned and headed away from their nest. With this simple but ingenious experiment, Santschi confirmed his discovery that ants used the angle of the sun to navigate.

Bert Hölldobler performed similar experiments on the harvester ants of the Arizona desert, but with some modifications. He caught ants that were carrying seeds back to their nest, which was located to the southwest, and released them 50 meters to the east. The ants kept going in the same direction according to the angle of the sun, but 50 meters to the east of their home. This added more credibility to the theory that ants use the sun as their compass.

Santschi's experiment proving that ants use the sun to navigate.

The results of Hölldob-
ler's experiment with
harvester ants.

Amazing Mathematical Ability of the
Sahara Desert Ants

Ants determine the angle of the sun in the sky as they go to where food is and then turn around 180 degrees to head back to the nest. It is truly amazing that these ants' tiny brains can determine the direction of the sun as well as how to turn exactly 180 degrees, but the story of the ants of the Sahara desert is even more amazing. Rüdiger Wehner of the University of Zurich researched this problem at length and he concluded that these ants did not simply turn around 180 degrees. Every time they changed direction, they measured the angle and "calculated" exactly how to go back to the nest from there in a straight line.

Chemical trails are useless in the constantly shifting sands of the desert, so these ants have to forage for food individually. They scramble around the desert looking for prey that have dried up in the blazing sun; anything they can eat, in fact. After scampering around in the sand, constantly changing direction, no matter how complicated the path to their food was, the moment they have it in their jaws they turn around and head back to the nest in a straight line. The geometric calculations these ants make from moment to moment would be extremely difficult for all but the most gifted of human beings, so how do these tiny creatures do it? Do they have miniature high-speed computers inside their tiny brains? Indeed, Rüdiger Wehner and his colleagues have discovered that the Sahara desert ant has a compass in the brain in addition to the leg odometer that measures the travel distance. But how such a small brain can analyze complex information on visual cues, distance data, orientation, and so forth is not known.

For that matter, how can forest-dwelling ants that see only flashes of sunlight through the canopy of trees find their way home? Bert Hölldobler found the answer to this question with ants that live in the African forest. To test his hypothesis that ants could remember the patterns of light and darkness made by the sunlight shining through the forest canopy, he performed a clever experiment. He made a large, dome-like glass cage for the ants he kept in his laboratory and then placed a picture of a forest canopy, taken using a fish-eye lens, over the top of the cage.

After he had given the ants time to acclimate to the picture of the forest canopy, the real experiment began. He waited until a worker ant discovered food on the side of the cage opposite from its nest, then while the ant was taking the food back to the nest he suddenly rotated the picture 180 degrees. After running around confused for a while, the ant headed in the exact opposite direction, back toward the place where the food was. When he turned the picture back around again, the ant changed direction again and headed back home. This experiment was similar to the methods Santschi used to discover how ants' sense of direction works.

Changes in the Sun's Position and Internal Clocks

There is a problem with using the sun to navigate. If a long time elapses between the time an ant leaves the nest looking for food and when it goes back to the nest, it will be difficult to use the sun to establish which direction to go in because the sun's position will have changed considerably. However, in experiments in which scientists found ants that were returning to the nest, caught them, and shut them in a dark box, the ants managed to accurately compensate for the change in the sun's position and find their way back to their nests. This was possible because a kind of internal clock inside the ants' brains told them that the hour that had elapsed while they were in the box meant they had to adjust for the fact that the sun's position had changed by 15 degrees per hour.

This phenomenon has been studied far more thoroughly in honeybees. If a scout bee that left the hive in the morning finds a particularly plentiful source of nectar, when it returns to the nest it will waggle dance for hours to let the other bees know where the source of that nectar is. Even though the bee cannot see the changes in the sun's position as it dances, every hour it will adjust the direction of the dance by 15 degrees to compensate for the change in the sun's position. Honeybee brains, too, contain a kind of internal clock.

Compared with ants and bees, human beings do not seem to have such ability to compensate for changes in the sun's position. We do, however, have a kind of internal clock of our own. People who take transoceanic flights may know in their minds what time it is in their new location, but their bodies stubbornly insist on keeping the same timetable they were on where they came from. Studies of people staying deep underground, with no exposure to the sun, have found that they still tend to live their daily lives on a 24-hour cycle. Strictly speaking, the average human body prefers to go by a 25-hour daily cycle. Were we born by mistake on a planet that revolves an hour too quickly? Perhaps in the ancient past the Earth rotated more slowly than it does now.

9

Ants at Work

maids, nannies, laborers, soldiers

Every ant is a member of a highly organized society, and every ant has its own job to do. The ants carry out their jobs faithfully. Once the queen has hatched enough workers to care for her, she dedicates herself entirely to laying eggs. Worker ants perform a much wider variety of tasks, beginning with the nanny work of caring for the queen, her eggs, larvae, and pupae, as well as seeing to odd jobs around the nest. As they get older they may move on to other tasks such as construction work, guard detail, foraging for food, and even going to war. They may start in the center of the nest, but as they grow older they gradually move outward.

Some ants rush into the workforce before they are adults; for example, the weaver ants that live in the treetops of Africa, Australia, and Southeast Asia and build their nests by sewing leaves together. When they build, workers use their jaws to pull together the edges of leaves from nearby branches while another worker fetches one of the colony's larvae and sews the seams together with strands of silk secreted by the larva when it is squeezed. The epitome of cooperation and sacrifice, the weaver ants mobilize their young as part of the workforce—a sort of child exploitation that in human society might land an employer in prison.

Different Species Have Different Jobs

Army ants are probably the best-known of the predatory ants, but there are also hunter ants that roam the wild individually, hunting for prey. Most of these ants belong to the subfamily Ponerinae, who kill their prey using poisonous stingers. One particularly remarkable ponerine ant is the trap-jaw ant of the genus *Odontomachus*, which is found throughout the rainforests of the world. This fearsome looking ant is a constant, lurking danger for small, soft-bodied insects like springtails, the trap-jaw ants' preferred prey. The trap-jaw ant holds its long, powerful jaws open almost in a straight line as it waits, completely still, for its prey to approach. Two pairs of long, thin sensory hairs protrude straight out from between the trap-jaw ant's open jaws—so fine that it takes careful observation to see them. Even the slightest touch to these sensory hairs triggers the jaws to snap shut, much like a bear trap. The sharp pain of the trap-jaw ant's bite may last only for a moment, but later the wound burns because immediately the prey is caught the ant stabs it with its stinger several times, injecting poison.

Bert Hölldobler and his colleagues from the University of Würzburg used a high-speed camera that filmed 3,000 frames per second to measure the closing speed of the jaws. They discovered that the ant's trap-jaw strike is the fastest-known animal movement in the world. Previously it was thought that the fastest move-

A yellow meadow ant (*Lasius flavus*) taking care of eggs.

A trap-jaw ant (*Odontomachus*) waiting for prey. If prey touches either of the long, slender hairs inside the ant's mouth, its jaws snap shut.

ment was the leap of a flea, which took about 0.7- to 1.2-thousandths of a second to complete. The trap-jaw ant's jaws close in as little as one- to three-thousandths of a second. According to these calculations, the tips of the trap-jaw ant's mouth move 8.5 meters per second. Proportioned for size, this is like a human being moving at 3 kilometers per second. That's faster than a bullet.

According to a recent study by Sheila Patek of the University of Massachusetts and her colleagues, the rapid strike of the trap-jaw ant has multiple functions. It is used not only in capturing prey but also in ejecting predator intruders and jumping to safety. The jaw strike, hitting the target, yields enormous propulsive force and instantaneously launches the ant backwards into the air, away from the predator. It is remarkable that a single mechanical system can produce profoundly different multiple behavioral outcomes.

As is well known, there are still communities of human beings in remote areas of Africa and Latin America who live almost entirely off of the fruits and seeds that they gather. Similarly, there

Opposite, A citronella ant (*Acanthomyops claviger*) taking care of larvae (*top*). A cornfield ant (*Lasius alienus*) taking care of pupae (*middle*). An Allegheny mound-building ant (*Formica exsectoides*) carrying gravel to help dig a tunnel leading out of the nest (*bottom*).

Top, A Costa Rican red ant harvesting the seed from a passion flower fruit. *Bottom*, A leaf-cutter ant returning to the nest with a leaf. The ants will grow mushrooms on it.

Opposite, A ponerine ant harvesting nectar from an extrafloral nectary to take back to the nest.

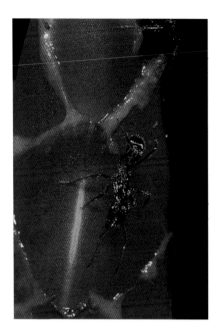

are gatherer societies in the ant world. The harvester ants of the Arizona desert each have individual territories where they gather seeds that have been carried there by the wind. These ants are highly competitive: they consume as much grain as do the local rodents. Among the gatherers are also the already-discussed leaf-cutter ants who grow mushrooms and the ants in many parts of the world that harvest nectar from plants or consume honeydew that aphids and other insects make for them out of juices sucked from plants.

When autumn comes in Korea, many people head out to the woods to gather acorns to make a tasty side dish called *dotori mook,* a kind of acorn jello. Ants join them in their search for acorns. but the ants are not gathering the acorns for food: they make their homes inside the acorns. These tiny ants, called *Leptothorax,* are not looking for fresh acorns but the ones that have been sitting on the ground since the previous year. This saves them some of the hard work of digging out the insides; they look for acorns that beetle larvae have already hollowed out. Some entire *Leptothorax* ant colonies can fit into a single acorn, but when the number of workers increases to the point that a single acorn is no longer big enough, they branch out into more acorns. There are believed to be eight different species of *Leptothorax* ants in Korea alone, but according to my observations they are becoming rarer and rarer. Signs are posted in many of the forests of Korea asking people not to take too many acorns—to leave some for the squirrels. Perhaps they should also mention the plight of the *Leptothorax* ants.

In many ant species, when pupae become adults they are not strong enough to themselves tear their way out of their cocoons: the worker ants have to tear open their cocoons for them. This is like being midwives at the births of their baby sisters. Aside from bats, which need help because mother bats have to give birth hanging upside down, and humans, ants are the only animals that have midwives to help them birth their young.

As in human society, there are many different kinds of jobs in the ant world. I have thought hard to name human occupations that the ants do not have. So far, I have found only two: the clergy and professorships.

Ants—Slacking on the Job?

Western culture has long considered the busy, eager beaver to be one of the hardest working animals. The dams that beavers build by cutting down large trees with their teeth are amazing feats of animal engineering. The expression "busy as a beaver" says it all. But ants work at least as hard as beavers. There is in fact an ancient Korean saying praising their industriousness: "Ants may be small, but they can build pagodas." Overall, however, ants turn out to be lazy. The colony as a whole may be busily moving about, but careful observation of individual workers reveals that they spend most of their day just sitting around. For that matter, beavers have been observed to work only an average of five hours a day.

Observations of Japanese carpenter ants, made in a laboratory by me and Dr. Sang-im Lee of Seoul National University, have

Top, A beetle larva has bored into this acorn. Once the beetle larva has hollowed it out, the acorn is ready for *Leptothorax* ants to move in. *Bottom,* An *Leptothorax* ant colony using a small acorn as a nest.

also shown that worker ants are active only 18% to 27% in any given day. Of course, a laboratory environment is more stable than the wild, and ants in the lab may have not as much work to do; they did, however, spend most of their time in one place, not moving. On surface this phenomenon appears to fit nicely into the so-called Pareto Principle, also known as the 80-20 rule. The Pareto Principle, named after Italian economist Vilfredo Pareto, was first suggested by a scholar of business management, Joseph Juran. It predicts that roughly 80% of effects come from 20% of the causes. In business economics, it supports a rule of thumb that 80% of sales come from 20% of the clients. In ant economics, however, are 80% of worker ants just sitting around doing nothing?

When we consider the colony as a superorganism, these worker ants are not so different from the cells of a human body with a healthy metabolism. Like many of our cells, which until they are needed are not always active, the majority of a colony's worker ants stay absolutely motionless, wasting little energy. In this way they can respond effectively to a crisis situation.

In the human situation, when a disaster like hurricane Katrina or the Fukushima tsunami strikes, the bulk of the rescue workers who flock to the scene are people who leave their jobs elsewhere to join the rescue effort. Perhaps because the history of ant society is so disaster-ridden, ants have evolved to maintain up to 80% of their workforce in reserve.

In the Old Testament, King Solomon exhorts, "Go to the ant thou sluggard; consider her ways, and be wise" (Proverbs 6:6). The truth of the matter, however, is that no animal works as hard as the human. The French, who have the shortest workweek of the industrialized world, punch in for an average of 1,600 working hours per year; workers in East Asian countries such as Japan and Korea clock up 2,000 or more working hours per year. These figures show that differences in working hours for humans is not so great, but compared with the animal realm it seems we have hardly any free time. Our washing machines, vacuum cleaners, computers, and other gadgets may have made modern life more convenient than that of our Stone Age ancestors, but they have not shortened our work day. In fact, our lives seem to become busier with every passing day.

Some animals store resources, such as food, for emergencies, but not enough to last longer than one winter at a time. Ants and humans are probably the only animals on the planet who compete fiercely enough for the resources around them to actually exhaust them—but ants are not foolish enough to work themselves to the point of exhaustion to trade in a car that works perfectly fine, just so they can buy the latest model.

10

Con Artists of the Ant World

*parasites that have cracked
the ants' secret code*

To human eyes, it really is quite a sight to see slave ants faithfully serving their slavemasters, even though the master ant looks nothing like the slave. Ants do have eyes and can see the differences between one object and another. However, since ants rely so heavily on a chemically based language, some things happen frequently in the ant world that we humans who rely so heavily on our sense of sight might find hard to believe. Humans can see that uninvited guests such as beetles and caterpillars that ants willingly give food to do not belong to the ant colonies they exploit; it is, however, their ability to imitate the ants' smells and behavior that cons the ants into giving them what they want.

Rove Beetles, Masters of Disguise

Rove beetles are the best-known of thousands of types of mites, moths, beetles, millipedes, crickets, and butterflies that live as the uninvited guests of ant colonies. When an ant nest is dug open, one can find rove beetle adults and larvae wandering around unmolested in the nursery, for example, or other heavily guarded areas of the nest. They receive food from the worker ants as if they were ant larvae, and they even feast upon the ants' larvae right before the workers' eyes. The worker ants go so far as to groom these con artist rove beetles as if they were their own larvae. How do the beetles pull off cracking the ants' code, slipping through their security net and swaggering around inside the belly of the beast?

The rove beetle larvae's technique of begging for food from passing ants by tapping on them is so refined that they do this in exactly the same way that ant larvae do. As soon as the ant's antennae feel the rove beetle larva's body, the ant lifts up the upper part of the larva's body and the larva feels around until it finds the ant's face. The larva then lightly taps on the ant's mouth, below the jaws, with its own mouth. The ant then feeds the larva by spitting up food into its mouth before carrying the larva to the nest. The rove beetles do not just give a password to get past the ant sentries; their trickery allows them to penetrate deep inside the enemy camp escorted by the very guards whose job is to keep them out. The rove beetle larvae are masters of disguise, with infiltration missions that are like something from *Mission Impossible.*

Rove beetle adults can also imitate the way that worker ants beg each other for food. The rove beetle uses its antennae to tap passing worker ants to get their attention, and if an ant comes over, the beetle taps on the ant's mouth with its front legs. The ant then spits up some food into its jaws for the rove beetle to slurp up. As if these mouth-to-mouth feedings were not enough pampering for the rove beetles, they even have glands that imitate the pheromones that ant larvae secrete to ask the worker ants to lick them clean. A rove beetle has secretory glands on both sides of its back that produce chemicals almost identical to the chemicals that the larvae are coated in so that the worker ants can identify them. The rove beetles' false chemicals are so effective that when they are squeezed onto a piece of paper shaped like an ant larva, ants will pick up the paper "larvae" and carry it to the colony's nursery.

Highwaymen of the Ant World

Not all rove beetles infiltrate the ants' nests in order to cause them trouble. Instead, they fleece the ants of the food that they are carrying back to their nests. Some rove beetles, also known as highwayman beetles, trick the ants by imitating the way worker ants beg each other for food. These beetles, however, are not equipped with the deceptive glands that other species of rove beetles have, so often the ants catch them at their game. When this happens, the highwayman beetle retracts its legs and antennae underneath its shell—like a turtle—and clings to the ground. Once the ants tire of trying to overturn the beetle, the beetle gets up as if nothing had happened and takes off to look for another sucker.

Some insects have graduated from victimizing the ants with petty larceny to killing them and sucking their blood—an act of outright robbery-homicide. The predatory insects of the Hemiptera order that are commonly known as assassin bugs kill and eat a wide variety of insects, and certain species of assassin bug specialize in preying on ants and termites. There is a kind of assassin bug that sits next to ant trails and tempts the passing ants with an irresistible scent that they secrete from a gland in their abdomens. When an ant approaches, drunk on the enchanting smell, the assassin bug lifts its legs to expose its secretory glands. The ant becomes so absorbed in lapping away at the assassin bug's secretions that it does not even notice the proboscis, or syringe-like probing tube, that the assassin bug plunges into the back of its neck. A few minutes later, the ant slurping away, paralysis sets in. The assassin bug stabs the ant with its proboscis to suck the blood from the limp, motionless body—as helpless as the pretty girls who swooned before the charms of Count Dracula before he sank his fangs into them.

Some assassin bugs use bait to lure their victims. These ingenious assassins start by casing a termite nest. They camouflage themselves with bits of dirt that the termites have dug out to build their nest and then approach the nest entryway. An assassin bug covered with such dirt smells like the surrounding soil. The assassin bug then catches a termite, injects the termite with poison via its proboscis, and sucks out the termite's blood. Next, the assassin bug approaches the nest with the freshly killed termite carcass on its back. The smell of the dead termite brings other termites from the colony, and the assassin bug uses each newly slaughtered carcass as fresh camouflage. Behavioral scientists are fascinated by the almost humanlike intelligence shown by the assassin bugs. How could such a tiny insect have come up with a tactic to camouflage itself and use the dead bodies of its prey as a lure?

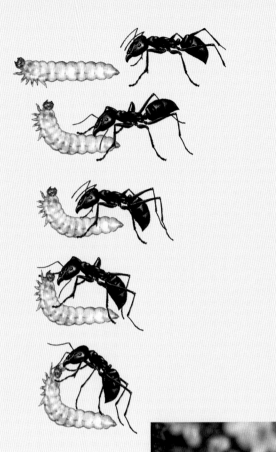

Above, An ant feeding a beetle larva. *Right,* A yellow meadow ant (*Lasius flavus*) taking care of larvae.

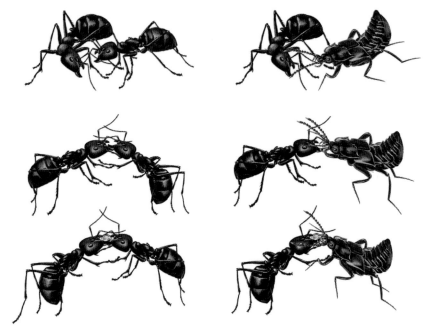

Rove beetles get food from worker ants by copying the behavior the ants use when sharing food. *Left,* Two ants share food. *Right,* Ant and beetle share food.

The Many Strategies of Lycaenid Butterflies

Caterpillars of lycaenid butterflies—one of the most diverse families of butterflies—have relationships with ants similar to that of the rove beetle. The caterpillars secrete appeasement pheromones that confuse ants so thoroughly that they even escort the caterpillars into the ant nurseries, where they feed the caterpillars all winter long. If only the grasshopper from "The Grasshopper and the Ant" had learned this trick! When spring comes, the caterpillar turns on the ants that have fed and carried it around all winter in the warmth of the nest. The ungrateful guest stuffs itself with ant larvae to prepare to enter its chrysalis stage. When summer comes, a magnificent butterfly emerges from the cocoon and flies away, never to return. Eventually, the lycaenid butterfly lays eggs on plants that its caterpillars will like, and the caterpillars that hatch will meet up with ants, and so it goes again.

Other species of lycaenid butterflies, too, have symbiotic relationships with ants. In fact, according to Naomi Pierce of Harvard University, more lycaenid butterflies live symbiotically with ants than live with them as parasites. The caterpillars from the former types provide honeydew for the ants in return for their protection from predators. The honeydew secreted by the lycaenid butterfly caterpillars of Australia provides their ant protectors with as many as 14 different essential amino acids, as well as many kinds of sugars. Evolutionary systematic analysis suggests that the ancestors of all lycaenid butterflies had symbiotic relationships with ants, but some species later evolved to have exploitative parasitic relationships with them instead.

A Costa Rican lycaenid butterfly with a false head. © Jae Choe

One of the remarkable aspects of the lycaenid butterflies that live symbiotically with ants is that in their adult form they, too, are masters of disguise. The pattern and shape of their wings confuse predators, making it hard for them to tell the butterfly's head from its tail. The stripes on the wings all converge in the wing tail, where there are structures that look like antennae. When these butterflies are at rest, they constantly wave these false antennae to add to the illusion. Some biologists have actually observed birds and lizards attacking these false heads—the butterflies flying away with nothing more than minor injuries to the tail ends of their wings. At some point in their evolution, they acquired the habit of deceiving others, not unlike their parasitic cousins.

11

Villains and Monsters
of the Ant World

predators and parasites

Many different insects fall prey to birds, lizards, spiders, and carnivorous plants such as sundews and Venus flytraps. But some predators specifically prey on ants. The larva of the antlion is an interesting example. In its adult stage, it is a flying insect that looks similar to a dragonfly, but it gets its name from its habit of viciously preying on ants during its larval stage. While assassin bugs—an ant predator that we met earlier—go to ant nests to prey on ants, antlions dig pit traps and wait for unwitting insects, especially ants, to fall in.

Antlion larvae dig funnel-shaped pit traps in sandy places and hide under the sand at the bottom to wait for their prey. The larva digs the walls of the pit at just the right angle. Once an ant makes the mistake of setting foot in the antlion's pit trap, it triggers a miniature landslide and the hapless ant falls to the bottom of the pit. The antlion, awaiting its victim, kicks sand away from the bottom, adding to the avalanche and sealing the ant's fate. No matter how hard the ant struggles, there is no escape. The antlion does not pounce on the ant as soon as it falls into the pit, preferring to wait until the ant is exhausted from its desperate attempts to escape. Then the antlion larva rears its massive-jawed head. According to Nick Gotelli of the University of Vermont there are places in Oklahoma that have clusters of hundreds and even thousands of antlion pit traps, each with an antlion larva waiting at the bottom. Places like this must be the stuff of ant nightmares.

A queen ant trapped on the leaf of sundew, an insectivorous plant.

A jumping spider feasting on an ant.

Opposite, A pit trap dug at the perfect angle. The antlion that dug the trap is waiting at the bottom for ants to fall in.

The Dragon of the Ant World

Ants in many different tropical regions of the world must live with the constant menace of several kinds of mammal that feast on ants. In Africa they have to contend with aardvarks (with a name like that, first in the dictionary); another species of African anteater uses its sturdy claws to dig up ant nests and then snake its long, sticky tongue deep into the tunnels, slurping up ants. Vested anteaters, named for the black fur that covers their shoulder and stomach areas like furry vests, live on Barro Colorado Island in Panama, where I did extensive field research. Whenever I ran into an anteater, the startled creature would rear back on its hind legs, brandishing its front claws and squealing at me, trying to look threatening. I could not help but find them cute and a little bit silly, although for ants anteaters are a force to be reckoned with.

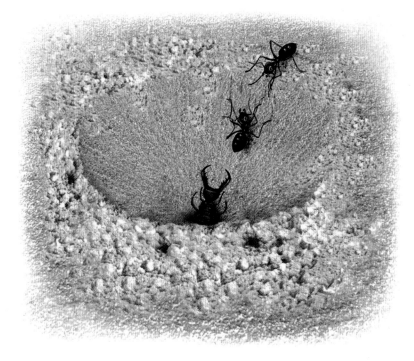

Most species of ants in those regions actually have more to fear from one of their own kind, however; namely, the army ants. While anteaters' tongues usually cannot reach the queen's chamber of most ant nests, if ants make the mistake of letting army ants into their queen's chamber or their nursery, the army ants will not hesitate to tear apart and consume everything they find there. I witnessed ruthless army ants invading many of the nests of other ant species. Each time I saw one of these attacks, I could not help but admire the way the invaded ants heroically held their ground to defend their colonies. Solzhenitsyn would have been impressed by their sacrificial acts.

The Porcupine of the Ant World

Beware the prickly porcupine! Many animals and humans have regretted making the mistake of disturbing one. The late Thomas Eisner of Cornell University has reported that there is an ant equivalent of the porcupine. This is a type of polychaete, or bristle worm, that defends itself from ants by squirming out of their grasp and covering the ants' bodies with bristles. The bristle worm's unusually sticky hairs restrict the ant's movement so much that such treatment can be fatal. The bristle worm is in fact a pricklier foe for ants than the porcupine is for larger animals since porcupines rarely kill with their quills.

Parasites

Soon after ant biologists began observing processions of leaf-cutter ants carrying leaves on which to grow their mushrooms, they noticed something that did not at first make sense. While one ant was carrying a leaf four or five times its own weight, smaller ants from the same colony were not only not helping to carry the leaf but were actually riding on top of the leaf. When I first saw this scene, I thought back to my childhood: as a child helping my uncles to harvest potatoes, I sometimes hitched a ride in one of their wheelbarrows; were these ants just getting a free ride for kicks? Field biologists—one of them being Don Feener of the University of Utah—then discovered that the small ants were not just playing

An Aztec ant nest on Barro Colorado Island, Panama. The nest is the long, thin structure on the tree trunk. Even though their nests, one to two meters long, are sturdy and built fairly high up in tree branches, Aztecs are not immune to occasional anteater attacks. © Jae Choe around while their big sisters did the heavy lifting. The hitchhikers have a job to do: they watch for parasites.

There are parasitic flies that take advantage of leaf-cutter ants while they are too busy hauling leaves to protect themselves. These nimble flies swoop on the preoccupied ants and lay their eggs at the base of the ants' heads, taking off again before the ants even know what hit them. Not long after, the fly eggs hatch and the maggots consume the worker's entire body. Needless to say, the leaf-cutter worker ants do not mind carrying the smaller workers along with their cargo in order to avoid this fate. But in spite of the ants' best efforts to chase these flies away, some ants fall victim to these parasites. The flies' eggs inside the victims' bodies hatch and the voracious maggots squirm through the ants' flesh.

Molds and Fungi

Although there are no statistics to prove it, it is likely that more ants fall victim to molds than to any other disease. Mold is a particularly serious problem for ants in the hot and humid tropics. During field work, I frequently saw ants and other insects covered with deadly molds.

Ants are afflicted by many different types of mold, from molds that look like fluffy coats to a mold of the genus *Cordyceps,* also known as the vegetable caterpillar. In Butan, China, and Korea, people actually cultivate cordyceps molds for their medicinal properties: they are useful in treating a variety of illnesses, including cancer. But to ants the caterpillar is deadly as its shoots sprout through a host ant's exoskeleton like thorns.

Mold is a constant problem in the humidity of the tropical rainforests. Sweat-soaked clothes left in the wrong place for even a day can be ruined by mold. In the tropical rainforests it is not uncommon for mold to even get inside the lens of a camera. At the tropical research station I worked at, each room had to be equipped with a drying booth, a closet with an incandescent light that was constantly left on to dry out whatever is put inside it to prevent mold. As soon as we went indoors, we put our cameras in the drying booth and hung freshly washed clothes inside the booth.

Above, Leaf-cutter ants hitching a ride. They do this to protect the colony's foragers from parasitic flies that lay their eggs on ants' necks. *Left,* Maggots that hatched from eggs inside an ant's body tearing their way out of its exoskeleton.

Mold in our cameras and clothes was the least of our problems. Despite our best efforts, like every biologist who has done field research in the tropics we ran into mold growing in patches on our own bodies. One of the more serious incidents I remember happened to Jonathan Coddington, a leading spider biologist from Harvard who is now a chief scientist at the Smithsonian Natural History Museum. When he returned to Harvard from Central America, a Harvard professor, Donald Pfister, one of the world's foremost authorities on molds, made Jon remove his shirts. At least five different species of molds were growing on his body. To this day, more than two decades later, I, too, find some molds growing back on my body during particularly humid summers. Some molds are tough to get rid of!

But what of the ants? Contagious disease is an inevitable consequence of being a social animal. Maintaining a low population density could wipe out the bacteria that cause diseases by depriving them of the opportunity to spread from one host to the next, but it is the very nature of social animals that has allowed them to dominate the Earth. This privilege, however, has clearly come at the price of disease. Although ants' hygiene methods have not been studied in detail, they seem to have surprisingly few contagious diseases considering the dense communities they live in, which suggests a high probability that over millions of years of evolution ants have developed methods of disease-prevention.

According to recent research, ants living in humid regions secrete a chemical that kills mold. They do this from glands in the sides of their thoraces. Interestingly, this antimold secretion has no effect on the mushrooms that the leaf-cutter ants grow: it kills only the mold and bacteria that could invade and harm the mushroom farms. A further recent discovery is to do with the chemical called propolis, a component of beeswax, which can be beneficial to cancer patients. The finding has caused some excitement, and ants are believed to manufacture similar substances. But this has yet to be thoroughly researched.

Some ants in the tropical rainforests of Central America line their nests with a kind of wallpaper to regulate the humidity. After the adult ants emerge from pupae, they hang the discarded cocoons on the walls of the nest. A research has revealed that the chambers that have been wallpapered with these cocoon scraps are considerably less humid than the chambers without this lining. Understandably, the first chamber of the nest to be papered is the nursery. These ants manifest their love for their young by hanging several layers of paper there to ensure that the next generation gets off to a healthy start in a clean, dry environment.

Top, A queen ant covered with a fluffy coat of mold. *Bottom,* A two-centimeter-long *Paraponera* worker ant with thorn-like shoots of cordyceps mold sticking out of her body.

A tropical jumping spider that has evolved to look like an ant.

Ant Mimics

Ants actually have relatively few natural enemies. Ants are the masters of their ecosystems, as we are of ours. It should come as no surprise that, again as with humans, the ants' greatest enemy is each other. Many species of ants, such as army ants, attack the colonies of other species, and there are also ants that seek colonies of their own species and devastate them, capturing slaves for their own colonies.

There are also many animals that mimic ants. Certain species of assassin bug have evolved a body with a slender waist that looks so much like the ants' that they can walk alongside ants without being detected. And in the Costa Rican rainforests I observed a species of treehopper that has a structure on its back that looks just like an ant. Unlike other species of treehopper that provide ants with honeydew in exchange for protection, these treehoppers are able to live independently, possibly because they can scare predators away by looking like ants. There are even species of predatory spider that mimic ants.

Nature is full of imposters. Among the most popular models for the imposters are bees, wasps, and ants. These three, along with termites, are known as the social insects. But why would other insects mimic primarily the social insects? It is probably because social insects are among the most fearsome. A single worker ant or bee may not pose a serious threat, but see one ant and you had better know that there usually is an army behind it.

PART III

The Politics of Ant Society

12

No Children of Their Own

I consider ants and humans to be the two rulers of the world. We run the civilized world because it is our invention. But if we step out of our world and move into the natural world, the ruler there is the insects, who ecologically are Earth's dominant creatures. And of all the insects, ants are arguably the most successful.

There could be many reasons behind the ecological successes of the ants and humans, but I want to single out one as the most important: cooperation. Our ancestors were able, through highly organized cooperation, to hunt down the massive mammoth, and ants, too, can cooperate and capture prey vastly bigger—a thousand times bigger, in the case of the ants—than their body size. If the premium of cooperation is this great, the question, then, becomes why every animal species has not learned to cooperate. Why don't mosquitoes cooperate? Why don't house mice cooperate?

Cooperation is not easy to achieve because in the process some members of the group have to sacrifice themselves. Democracy is a system in which the individual members constantly try to even out the per capita sacrifice among themselves, but this is a nearly impossible goal. Self-sacrifice or altruism is such a difficult virtue that only a handful of animal species have attained it in the history of evolution. Ant colonies are often compared with the Amazon tribe of warrior women from Greek mythology because, except during a brief mating period, they are composed entirely of females. Along with the queen, all of the workers responsible for work done both inside and outside of the nest are female.

Pheidole worker ants cooperate in attacking a large wasp.

Ant males lead relatively short lives All born during the mating season, they leave the nest together on a warm, tranquil day for a fleeting tryst with eligible females from other colonies, then die shortly afterwards.

That ant colonies are almost exclusively composed of females has proved to be hard for many people to swallow. When artistic creations depict ants they are often either mistaken or intentionally inaccurate. The American computer-animated action adventure film *Antz*, released in 1998 by DreamWorks Animation, featured as the main character a male worker ant named z-4195. Science-fiction novels and movies often come with fictitious setups, but in the case of *Antz* this central aspect of the story was simply wrong. Perhaps the oldest literary work that involves the description of ant societies is *Nanketaishouzhuan*, or *The Governor of Nanke*, a Chinese novel from the Tang dynasty, written by Li Gongzuo (778–848). In his daydream, Chunyu Fen is invited to the kingdom of Ashendon, located underneath a huge ash tree next to his house. Having lived a successful but tumultuous life in Ashendon, he returned home and realized that the day on which he fell asleep had not yet ended. He uncovered a huge ants' nest under the ash tree and learned that his dream life was unfolding in the country of ants. The moral of the novel was the emptiness of materialistic life, but the ant society described in the novel was again far from the facts. Ashendon was just like a human society, with males and females, both having their own jobs to do.

Ant colonies have also been compared to nations organized around a single queen. Strictly speaking, however, it would be more accurate to call an ant colony a family. Aside from slave ants that have been kidnapped from other nests, every colony's workers are the daughters of the same queen. In other words, each colony is a family consisting of a widow with many daughters. The queen may have sons sometimes, but once they are married off the family goes on and once again only the daughters remain. When the sons are living at home, they never hunt or work around the nest; all they do is wait for the day of the nuptial flight. By human standards, ant males can only be described as slackers.

The daughters, however, lead very different lives. They never marry and devote themselves entirely to taking care of their mothers. Of course, a few of the daughters will leave the nest for their nuptial flight with the males, so they can start their own colonies as the next generation of queens. Most of the daughters, however, will spend all of their days working tirelessly to care for their mother and siblings. Their work begins on the day they are born, when they wait on the queen, and as they grow they move on to taking care of the eggs and larvae, until they are old enough to leave the nest to forage for food or guard the nest.

The Workers' Noble Sacrifice

It is remarkable that worker ants are willing to dedicate their entire lives to working for the colony, and even more remarkable are some of tasks they undertake. Some of their efforts stand out as particularly unusual. Honeypot ants of the deserts of the American southwest and Australia do a head-over-heels stint in service of the colony. The honeypots are predators, but they also protect insects such as aphids in order to gather and store the "honey" they produce for times when food is scarce. The "pots" they use to store the honeydew are large worker honeypot ants who hang upside down from the ceiling of the nests. Other ants from the colony feed them honeydew and they store it inside their stomachs. The honey load must be heavy, but these living honeypots hang for days and weeks for their colony.

Some ants function as doors. This can be seen in turtle ants and European carpenter ants. These ants live inside the holes they bore into large trees, and some of the workers, who have heads of an unusual shape, stand guard by using their heads to block the nest entryways. When workers from their own colony return to the nest, they tap on the sentry's head with their antennae and the sentry steps aside to let them into the nest. They will not move aside for any other colony's workers, no matter how much they may tap. Li Bai, one of the greatest poets during China's Tang dynasty, recited, "Thousands of soldiers may not open the gate held fast by a single gatekeeper," and, like the human guard, these gatekeeper ants will strive to protect their fortresses against impossible odds. Groups of these ants will literally put their heads together to protect an entryway if it is too large for one head to block. These worker ants begin their guard duty almost as soon as they are born and continue until they die. Their unflinching sacrifice in defense of the colony is astounding.

Some human beings, of course, make the ultimate sacrifice. Japanese kamikaze pilots of the Second World War, for example, crashed their planes into U.S. aircraft carriers, sacrificing their lives for their country. An impressive and not dissimilar example of this can be found in ants of the carpenter species of Malaysia. The worker ants of this species could even be called walking bombs. Each of these ants has a pair of long glands that run from their jaws to the tips of their abdomens. If the ant is attacked by a predator or during war with other ants, these glands explode, covering the enemy with a sticky poison and also killing the ant.

Carpenter ants use their unusually shaped heads to block
the entrances to their tunnels.
Top left shows the shape of the
head. *Top right* shows how the
head fills the entrance. *Middle
and bottom*, Carpenter ants
guarding an entryway. These
ants will move aside only when
a member of their own colony
taps on their heads.

If a honeybee has ever stung you, it did more than just inject you with venom. Before flying away, the honeybee leaves its entire stinger and venom sac in the sting target. Afterwards, the honeybee will die, within about two hours. The muscles in the stinger continue pumping venom into the enemy body even after the stinger has been separated from the honeybee. In addition, if the event happened near the bee's nest, in defense of the hive, the venom evaporating from the other end of the stinger alerts other honeybees to the presence of an enemy so that other bees from the nest can swarm and attack.

For years, Americans have been in a buzz about a threat from killer bees, fearing that as they come north from the tropical regions of Latin America there will be more human fatalities. In fact, the real terror of these killer bees is not the number of people that they kill but their ferociousness when they do kill human beings. I know of a horrific incident that happened in the Palo Verde region of Costa Rica in the summer of 1986. A graduate exchange student from Malaysia, studying through the University of Florida, was killed by killer bees while climbing a mountain. He tripped over a crack between two rocks, disturbing a nearby nest of killer bees. The student's foot was stuck in the crack and he could not pull it out in time in spite of his fellow students' attempts to rescue him. He became unconscious and died after hours of screaming for help.

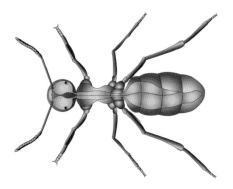

A kamikaze carpenter ant. These ants explode their bodies to cover their enemies with venom.

Genetically, killer bees and ordinary honeybees are similar. The difference is in their determination to protect their hives. Killer bees are much more sensitive to outside stimuli than ordinary honeybees, and the reaction they show to such stimuli is much more intense. In other words, these bees are more determined than most to sacrifice themselves. Due to their paranoid behavior, they are also able to produce more honey than ordinary honeybees, which is why they were transported from Africa to Brazil in the first place. Ever since they have gradually spread, now having reached as far north as the United States.

Dedicated to Working, No Offspring of Their Own

Worker ants and bees sacrifice more than their lives. In their evolutionary history, they have given up the ability to give birth to their own young. They have done this so that they can dedicate their lives to helping their queen handle all of the reproduction for the entire colony.

According to a theory of evolution that is based on the idea that all organisms exist in order to pass their genes on to the next generation, it should be impossible for any organism to make this sacrifice. In evolutionary terms, for an organism to commit to living its entire life without passing on its genes is tantamount to dying. How, then, in an animal kingdom that is fundamentally selfish, did these altruistic beings come into existence?

Charles Darwin, who proposed the theory of evolution through natural selection, was troubled by two things. One was the question of why, among all of the species that have two sexes, the males usually have more splendid characteristics. The other was how worker ants and bees could have evolved to sacrifice themselves for others in the colony. Both of these phenomena appear to be violations of the theory of evolution that all organisms adapt to their environments in order to survive so that they can propagate future generations.

The sacrifices of the worker castes of social insects—namely, bees, wasps, ants, and termites—form the foundation of their or-

A termite queen, swollen with eggs, destined to spend her entire life laying eggs and being fed by her workers.

derly societies. With the exception of termites, all social insects belong to the Hymenopteran order of insects. Hymenopteran insects have an unusual means of determining which of their offspring will be male and which will be female. Queen ants and bees have an organ called a spermatheca, used to store the sperm they receive from the males during their nuptial flight. Later, when the queens lay their eggs they will use the sperm that they have stored to fertilize some of the eggs in order to give birth to females and will close off the spermatheca to leave other eggs unfertilized in order to give birth to males. It can be said that ant and bee males are born from a virgin.

Most animals, including humans, are diploid, which means that each of their cells contains two sets of chromosomes. When they make sperm and eggs, however, each sperm and egg contains only one chromosome set: when they are combined in the form of a fertilized egg, the animal will then be born with two sets of chromosomes. Males of bees and ants, however, are born from unfertilized eggs; that is, they are born as haploid animals, which means that each of their cells contains only a single set of chromosomes. As strange as it may seem, ant and bee males do not have fathers. A male has a grandfather on his mother's side, but no father or paternal grandfather.

We humans, the self-proclaimed supreme species on Earth, having developed a highly organized civilization, with a dominant role in our ecosystem, are also social animals. By forming societies and cooperating with one another we have won the competition with the other species. However, during this process there has always been a conflict between the haves and the have-nots. How does ant society compare?

All ants in a colony are born from the same mother. Some ants become queens, while others end up as workers. At first glance, ant society seems to be unfair to the female ants that are born as workers. Yet, when I look at the queen ants—so bloated that they cannot move comfortably; spending their entire lives having to live on what the workers bring them; doing nothing but lay eggs— I cannot help but wonder who is really using who in ant society. Sometimes I find myself comparing queen ants to the surrogate wives of ancient times or the concubines that noblemen secretly kept in their homes when their wives were infertile. Much like ant queens, these women were shut away, swollen with child, dedicated solely to giving birth.

13

Conflict in the Queendom

divine right of the queen?
or the will of the masses?

When I observe ant colonies, these stiflingly organized societies with every individual faithfully performing its duty often remind me of George Orwell's classic novel *1984*. In human society, such events may largely be fiction, but ant societies have to some extent made them a reality. In ant colonies the life of the individual seems subordinate to the common good. The individual worker must sacrifice her own reproductive capacity so she can devote herself to helping her mother, the queen, handle all of the reproduction. The sacrifice may seem altruistic, but when analyzed genetically this is actually selfish behavior.

The Gene's Point of View

Genetically, how similar are we to our mothers? That is, how many genes do we share in common? The answer is that every human being has exactly 50% of the mother's genes. This is because one of the two sets of chromosomes comes from the egg that the mother provides at conception. The other 50% of a human's genes come from the father. When our bodies make reproductive cells, sperm or eggs, through the process of meiosis, we impart only half of the chromosomes that are in the somatic cells of the rest of our bodies.

Another genetics question: how much genetic similarity is there between brothers and sisters? Brothers and sisters who share the same mother and father have on average 50% of genes in common, that is, a genetic relationship of 50%. Even though two brothers may inherit the exact same genes from their mother, it is possible that one brother will take after the mother and the other take after the father. Hence, the average genetic relationship between two brothers is estimated to be 50%. The same is true for sisters.

The more distantly two people are related, the lower the genetic relationship is between them. British geneticist J. B. S. Haldane quipped at his favorite pub, "I would be prepared to lay down my life for two brothers or eight cousins." He was meaningfully referring to the fact that we share one-half of our genes with our brothers and one-eighth with our cousins. With aunts and uncles, we share one-fourth of our genes.

However, Hymenopterans such as ants and bees have more unusual genetic relationships between individuals due to their peculiar sex-determination system. Unlike diploid organisms such as we humans, they are haplodiploid organisms; that is, males carry only one set of chromosomes, while females carry two. According to British biologist William Hamilton, who established the theoretical foundation to answer this question in 1964, worker ants and later generations of queens born from the same mother have genetic relationships of 75%—much higher than the 50% relationship found in brothers and sisters of diploid animals. This high genetic relationship occurs because worker ants' fathers only have one set of chromosomes to pass on through their sperm. Ant males pass all of their chromosomes to their daughters. Although ant males die before they can see their daughters, they give daughters everything they have and thus have a 100% genetic relationship with them.

The 75% genetic relationship between worker ants is much higher than the 50% relationship they would have with their own offspring, were they able to have them. For this reason, Hamilton suggests we should consider the situation of worker ants from a genetic viewpoint rather than an individualistic viewpoint. If we look at evolution as the struggle to pass as many of one's genes as possible on to the next generation, then worker ants are doing a much better job by helping their queen to produce more workers and the next generation of queens. As they share a 75% genetic relationship with their younger sisters, helping to raise them does more to ensure the survival of their own genes than it would for them to give birth to and raise their own offspring, with whom they would only have a 50% relationship. Hamilton maintains that this is why worker ants have evolved as altruists, abandoning their own reproduction in order to dedicate their lives to their queen.

Oxford biologist Richard Dawkins's book *The Selfish Gene* posits that the basic unit of evolution is not the individual organism, but rather the gene. The individual is born, lives, and dies, but the gene is capable of living forever through reproduction. Explaining this, sociobiologist E. O. Wilson cites Samuel Butler's famous aphorism, "A hen is only an egg's way of making another egg."

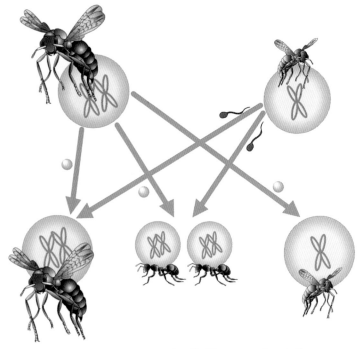

Genetic relationships in the ant family. The queen (*top left*) passes on 50% of her genes to her offspring (*bottom row*). The male (*top right*) passes on 100% of his genes to new queens (*bottom left*) and worker ants (*bottom middle*). Male offspring (*bottom right*) are produced solely by the queen.

Is the hen the egg's way of making more eggs?

Rebellion among the Worker Ants

Ant and bee colonies are often referred to as superorganisms because the individual members of these colonies have such strictly delineated roles to play: they cooperate so effectively with each other that their behavior is more reminiscent of the cells of an organism than that of independent organisms. The only job that queen ants do is lay eggs; they depend on the protection and support of the workers, and the workers are responsible for all of the work in the colony so that the queen can dedicate herself exclusively to laying eggs. This strict adherence to this division of labor maximizes the efficiency of each colony.

However, worker ants are not completely devoid of individuality like most of the human characters in the novel *1984*. They are still living beings with independent bodies and individual lives to live. Worker ants are not physiologically incapable of laying eggs but they are prevented from developing eggs by the suppressing effects of a pheromone the queen produces. This so-called "queen substance" is the medium that keeps the colony's division of labor in balance. Queen ants of course lay eggs; but there have also been many observed cases of worker ants laying eggs. When I was studying at Harvard in the mid-1980s, in a review article I listed all the research papers that reported instances of worker ants laying eggs. From this list of research papers I found that worker ants lay eggs quite frequently and that they do so for many different reasons. Worker ants lay two different kinds of eggs. One type is known as trophic, or nutritive eggs, which will not hatch into larvae but will serve as food for other members of the colony. The other type, reproductive eggs, hatch normally. Unless the worker is able to mate with a male, however, the reproductive eggs she lays will be unfertilized and the larvae that hatch from the eggs can grow up only to be males. In many ant species, if an ant queen dies, the remaining worker ants will begin laying eggs once their reproductive capabilities are no longer suppressed by queen substance. There are also ant species, however, whose workers lay eggs while the queen is still alive, and they care for the males that hatch from these eggs just as they would care for other males. Japanese and French biologists have also discovered that in at least two species of ants, workers lay unfertilized eggs that hatch into workers, not males. These unusual ants have somehow broken the rules of the haplodiploid sex-determination process.

The queen does not always forbid her workers from raising their own males. Certain species of ants have even evolved into a kind of "truce" between the queen and workers, and these workers produce some or even all of the males for the colony. In many species, however, there is a never-ending conflict between the queens and workers. Workers often clandestinely lay eggs on the outskirts of their colonies or barricaded in their own chambers within the nest. It can also happen in colonies where the queen substance that suppresses the workers' reproductive processes have suddenly weakened. They say you cannot outrun the long arm of the law, but there are limits to a queen ant's reach.

In some cases, when a queen ant discovers this kind of rebellion in her colony, she will hunt down the rebels, find their lair, and tear them apart with her strong jaws. These worker ants refuse to blindly sacrifice themselves for the greater good of the colony, in-

A worker ant taking care of eggs. In some colonies, worker ants lay eggs.

stead carving out their own niches out of sight of the queen's tyranny. They act in their own interests. These worker ants struggle to secure their own direct reproduction because they, as well as the queens who keep them in check, are still individual living beings. When I think about these ant rebels, the foolhardy and individualistic hero z-4195 of the animated movie *Antz,* fretting about his own needs in the irrational system of the ant colony, does not seem so farfetched. Protagonist z (his short-form nickname) is an individualistic worker who longs to make a life of his own in a hopelessly conformist society.

Social Conflict and Strengthening the Power of the Queen

Every worker ant and male ant is equally precious to the queen. Whether she fertilizes them or not, every male and worker is her son or daughter to whom she has passed on one-half of her genes. When establishing her new colony, she cannot of course begin with her sons—because males do not help with the housework. If we consider the ant queens and the males—the two elements responsible for producing future generations—to be the reproductive cells of the ant superorganism, then the worker ants who make it possible for the queens and males to grow up to reproduce would be the somatic cells. In other words, in a manner similar to how humans must develop tissues and organs in order to grow into adults who can produce the next generation, ant colonies must invest in workers so that their males and new queens can breed. Once a colony has developed an economic structure strong enough to produce males and new queens, the queen may begin investing equally in sons and daughters.

The workers, however, have a different agenda. As we learned above, worker ants from the same parents have a 75% genetic relationship with each other but only a 25% genetic relationship with their colony's males—their brothers. In diploid animals, brothers and sisters with both parents in common an average share 50% of the genes, but this is not the case with ants. Ant fathers do not pass any of their genes on to their sons, so worker ants have only a mother in common with their brothers, which means that they share only 25% of their genes. Therefore, contrary to the queen's way, worker ants will favor the new generations of queens and workers over the colony's males. While the queen may lay the eggs, it is the workers that take care of them. No matter how powerful a monarch's army may be, no monarch can rule without the support of the subjects. For ants, the result is an epic struggle between the queen and her daughters for control of the queendom.

Who has emerged victorious in this ancient struggle for power? According to most myrmecologists, it is the queen. There are ant species with queens and workers that have bodies of similar sizes, but even in those cases the queens have larger abdomens and the remnants of their wings. Such species are not the norm, however; in most ant species the queen is a great deal larger than the workers. They are also considerably stronger and have special abilities that allow them to control their colonies. The pheromones that the queen secretes not only allow her to suppress the reproductive functions of the workers, they also function as the source of her authority.

For an example of the power of queen substance, consider the slave-makers. Slave-maker ants kidnap larvae of other species' ant colonies and daub some of their own queen ant's pheromones on these larvae to raise them as slaves. The dollop of queen substance is enough to compel the abducted slave ants to spend their entire lives faithfully serving their mother's murderers as they would serve their own mothers. And as we learned above, in most other ant species these pheromones are also strong enough to prevent most worker ants in the colony from laying eggs.

To make her colony strong the queen must produce as many workers as possible, but she cannot do so unless she has sufficient quantities of sperm. There are limits, however, to the amount of sperm she can get from a single male. This means that most queen ants must mate with multiple males during their nuptial flights, obtaining enough sperm to last them their entire lives. Queen ants of course have no moral scruples about having relations with multiple males, but multiple mating can still lead to problems within the social structure of the colony. Although worker ants with the same mother and father may share a 75% genetic relationship, the many workers that are only half sisters because they have different fathers would not even share a 50% genetic relationship. In these cases, it would be to the worker ants' advantage to have their own offspring instead of helping the queen by taking care of their sisters. Nevertheless, the fact that so few of these half-sister workers rebel reproductively is evidence that monarchism has become firmly established over the course of the ants' long history.

A leaf-cutter queen attended by worker ants. The size difference is enormous.

Edward O. Wilson has recently taken issue with the limited prowess of a "gene's-eye" view of the evolution of social behavior. One of the many problems he pointed out was the inflated emphasis of haplodiploid sex determination and reduced genetic relationships due to the queen ant's multiple mates. Therefore, he denounced Hamilton's theory and reinstated group-level selection as the far more potent evolutionary force. This sudden change in his stance stirred heated debates among evolutionary biologists. Being at the center of controversy is not at all new to him and he is putting up another tough fight as we speak.

14

Political Conflict and International Alliances

yesterday's comrade,
today's enemy

In human history, some countries have always been ruled by a king or a queen—an absolute monarch. Throughout such countries' histories, none of them has had more than one king or queen at any given time. There are also many modern democracies whose presidents or prime ministers hold absolute power, even though their constitutions, on paper, may delineate clear separation of powers. Similarly, in most ant monarchies, a single queen holds power. With the exception of slave-owning colonies, all of the queen's subjects are genetically related—all the offspring of the same queen. In ant societies, this genetic relationship forms the basis for a level of self-sacrifice and cooperation that would be difficult even to imagine in human society.

Some ant species, however, form colonies ruled by multiple queens. This occurs most frequently in recently founded ant colonies. Most queen ants establish their colonies alone, but it is not uncommon for several queen ants to band together when they form new colonies. Such colonies, founded by multiple queens, are one of my main research subjects.

Troop Reinforcements Are Top Priority

The overwhelming majority of ant colonies are ruled by only one queen, however, including those that were founded initially by multiple queens. Most colonies ruled by multiple queens during their early stages do not maintain this system as they develop. Several queens may work together at first, but usually a single queen eventually seizes control of the colony. Although it may be advantageous for a queen to

Worker ants removing a young queen that tried to establish a new colony in their territory.

join forces with as many other queens as possible during the early stages of the colony, this will also decrease her chances of later seizing control of the colony for herself. Why, then, do queens come together if it inevitably leads to a bloody struggle for power?

Founding a new nation is a difficult task, in both the ant world and the human world. A new ant queendom begins once the queen completes her nuptial flight and picks a place to establish her new colony. She breaks off her wings, which she will never need again, and digs out the chamber that will be her new home and her throne. As soon as she settles into her new home, her race against time begins. Giving birth to and raising her young alone, cut off from the outside world, is difficult, but the queens probably evolved this survival technique in order to protect themselves from predators. The only source of energy she has at this point is what she can break down from the fat under the skin and muscle tissue where her wings used to be. Hidden away in her tiny room, the queen must produce enough workers to bring her food from the outside before her limited resources run out. If the queen fails in this battle against time, the colony will die before it has hardly begun.

There are few places on Earth where ants do not live. They live in the back alleys of our metropolises; they make their way into our homes. Ecologically speaking, ants are among the most successful animals in the world. The ants' triumph, however, is the result of a brutal and bloody struggle. There is never enough land to go around for new colonies to be founded in, so countless queen ants have died trying to establish their colonies in rival territory. Since there is a land shortage, the small spaces that are left between larger, already established ant colonies often have many queens crowded into them, each one rushing to found her own colony. Most colonies founded by multiple queens can be found in areas where competition is particularly fierce. In these conditions, with new queendoms sprouting like mushrooms after a rainstorm, the only way for a queen to survive is to build a stronger army, faster than any other queen.

Ancient Chinese historical texts tell the story of the revenge of King Gou Jian of Yue and how he vindicated his kingdom's humiliating defeat at the hands of the state of Wu through a complex plan to build up his armies. Over a period of ten years, he punished the parents of unmarried men and women of marriageable age and forbade all marriages between younger men and older women in order to increase the population of Yue. He also offered many incentives to parents who gave birth to many children. The population grew, King Gou Jian instituted a system of strict military discipline to train his armies, and he eventually drove out the armies of Wu. Ants, too, have to increase their populations in order to strengthen their armies, but a queen can produce only so many workers during the early stages of her colony. Several queens laying eggs together, however, can produce many more workers. Research has also shown that when queen ants cooperate, it decreases the amount of time it takes them to raise new workers.

The production of a larger army in a shorter time allows these colonies to crush the smaller colonies around them. If they did not cooperate, these queen ants would not be able to found even small communities, much less great nations. British poet Samuel Johnson once remarked that it is mutual cowardice that keeps us in peace.

Alliances Transcending Skin Color

Ant nations, much like human nations, usually come together because their members share a common bloodline. However, instances of interspecific cooperation have been discovered. An example of this comes from two species of Aztec ant that Dan Perlman of Brandeis University and I have been studying since 1984. The Aztec ants get their name from the once great Central American Indian civilization whose empire fell before the might of the Spanish conquistadors in the sixteenth century. These ants belong to the genus *Azteca* and inhabit the tropical rainforests of Central America. Most species of Aztec ants make their homes in nests hanging on long tree branches, but some species have a symbiotic relationship with trumpet trees of the genus *Cecropia*.

Trumpet trees have hollow trunks, much like bamboo trees, and Aztec ants make homes in the cavities. Even more remarkably, trumpet trees produce nutrient-rich food packets called Müllerian bodies for the ants that live on them. In return for the food and shelter that the trumpet trees provide, the Aztec ants patrol the trumpet trees to protect them from herbivores that might want to eat them.

The full-grown trumpet trees can reach as high as the canopies of the tropical rainforests. They are also the first plants to sprout when sunlight breaks through the dense jungle canopy after a large tree falls. As a trumpet tree grows upward, the Aztec ant colonies inside the tree grow along with it. Look at a vertical cross-section of a young trumpet tree and you will see that it is divided into sections—looking rather like the floors of a high-rise apartment building. Each floor has a different family living on it.

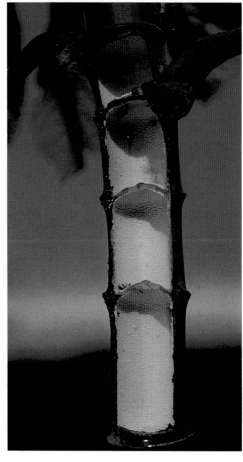

Top, Trumpet trees like this one have hollow trunks in which ants can nest. *Bottom,* Such nodes in trumpet trees provide shelter for Aztec ants.

Top, An Aztec worker ant gathering Müllerian bodies. *Bottom,* Aztec worker ants on patrol on a trumpet tree.

These new nations being formed inside the trumpet tree may be founded by single queens going it alone, but they may also be formed by multiple orange- or black-skinned queens pooling their strength to form confederacies. Just as our modern-day multinational corporations have learned to transcend national boundaries and join forces with one-time mortal enemies, so have the Aztec ants learned to transcend the bloodlines of their species to form multinational confederacies. Even though black-colored *Azteca constructor* and orange-colored *Azteca xanthacroa* ants are completely different species, their queens can be found setting their differences aside and working together much like the models in a United Colors of Benetton ad, forming new colonies.

However, the honeymoon is short for these marriages of convenience. As they give birth to their worker ants, the relationship that transcended interspecific boundaries to bring the queens together subsides and the inevitable, bloody struggle for the throne that is only big enough for a single queen begins. When the tree is young and small, it may contain several ant colonies led by singular or multiple queens, but by the time the tree is full-grown it will almost inevitably be owned by a single colony led by one queen.

An old saying goes, "Keep your friends close, but your enemies closer." When an ant queen is at her weakest, she may raise her young with other queens for the sake of her own survival, but eventually these other queens become enemies sharing the food that her own workers bring into the nest. Aztec queens will fight each other for power, but in most other ant species the queen of the empire is determined by "people power," so to speak. In a majority of ant species with multiple founding queens, the workers choose a single queen. As unbelievable as it may seem, during this process of selecting a queen, some worker ants will even kill their own mother on behalf of their new candidate for queen. In these ant societies, flesh and blood loyalty appears to play no part in determining the future outcome of the colony. Instead, the workers make a calculated decision based on which queen is carrying the most eggs. These ant societies are as ruthless as they are rational.

Queen ants forming new colonies in each level of a trumpet tree.

Top, An Aztec ant colony before the queens' bloody battle. *Middle,* Ant queens working together as partners before workers bring them food. *Bottom,* Suddenly the queens turn on each other in a fight to the death.

15

The Foundation Myth
of the Aztec Ant Queendom

. . . .

the queens' battle for the throne

The histories of human nations contain bloody struggles for power, and ant societies are no different. New ant colonies are constantly sprouting up, even in the tiniest slivers of land between firmly established, large, powerful ant colonies. More often than not, their only chance for survival in these fiercely competitive conditions is to band together. Colonies that try to go it alone out of xenophobia will not last long. To have hope of survival, they must overcome their differences of skin color and national boundaries.

In the summer of 1984, I was studying the ants that lived in the trumpet trees of the Monteverde cloud forest of Costa Rica to see if the stories I had heard about ant queens of different species cooperating with each other were true. Wouldn't each queen in the spaces between the cells of the stems of the hollow, bamboo-like trumpet trees found a colony of her own? Some of these ant queens were going it alone, while others were teaming up with other queens—from their own species and even queens from other species—to found new colonies.

One year later, Dan Perlman, working on his PhD, decided to study the Aztec ants as they founded their colonies in the wild. He was also interested in their subsequent struggles for power. I had already begun my own doctoral dissertation on a different subject, but I had not lost—and have still not lost—my interest in those topics of research. While I traveled around Costa Rica and Panama doing my own research, Dan settled in Monteverde with his wife, Nora, for more than a year, focused exclusively on studying the Aztec ants. I visited him there to work together on various projects during those days. Since then, we take turns visiting our field sites for follow-up studies. Although he has had a much greater share of the work, Dan and I have maintained a beautiful partnership to this day.

The Heroic Tales behind the Fun

This may be true of other fields as well, but the study of animal behavior sounds exciting when you hear the stories of the discoveries that scientists make. These discoveries, however, require roughing it in the wild for long periods of time without the comforts of civilization. I have known many people who rushed into the field, excited by the amazing animals they had seen on TV, only to move on to other things because they could not handle the hardships that the research requires. The most important qualities for any student of animal behavior are the willpower to endure the hardships of wilderness life and the patience to watch, wait, and observe for long periods of time without getting bored.

As a child, I always preferred playing in the mountains and fields to studying books, but even so I sometimes find it difficult to endure long periods of time living out in the boondocks. For me, the worst part is having to try to sleep on the moist beds that are an inevitable part of life in the humid rainforests. It is not easy to fall asleep when the air is humid enough to soak through your underwear, all the way to the skin. It brings back my memories of waking up after wetting the bed as a child and not being able to fall asleep again. That makes it even harder to bear. I hope that anyone thinking about going into the field of animal behavior considers, before beginning his or her studies, the conditions that have to be endured during field research.

The most important part of successful research in animal behavior is making sure that all observations and experiments are carried out without disturbing the subject animals' natural lives and habitats. In the mid-twentieth century, European scientists developed ethology, a branch of behavioral science that emphasized this principle of noninterference, thereby marking a new beginning for the study of animal behavior. Dan and I decided that we wanted to observe the foundation of an Aztec ant colony starting from the moment the queen lands on her trumpet tree, and we developed a long-term plan to make sure we observed them doing so in their natural habitat with little interference.

With the steam rising from the ground after a rain shower, an ant queen lands on a small trumpet tree after her nuptial flight. She flutters up and down the tree trunk a few times before ripping off her wings and starting to burrow her way into the tree trunk. She does not bore into just any spot on the tree. The tree has already provided for her a thin spot on each of its trunk segments. Even though the plant has done this for the queen, it still takes anywhere from 30 minutes to two hours for her to dig the hole and wriggle her plump body through it. Once she has made it inside, she scratches the inner wall of the tree trunk to produce material that will plug the hole she made.

Other queens usually already occupy these high-rise apartment buildings by the time a new queen arrives, although this might not be the case with a very young sapling. Once the first queen to enter a tree has made and closed her crude front door behind her, the wall of the tree trunk will grow back in that spot even thicker than the walls of the rest of the trunk. The door is "locked." From then on, new queens cannot get inside.

In older trees, queens who moved in earlier are now in the lower floors, having lived behind locked doors for some time. The queens on the midlevel floors may have closed their doors behind them when they moved in, but the doors might not yet be locked. Even though the first queen to go into a tree at this level may have blocked the hole she made, it would not be difficult for a newly arrived queen to push her way inside. According to our observations, it takes between five and twelve minutes to do so. This leaves a newly arrived queen ant with a choice to make. The lower floors are sealed shut, so she can either choose to move into the midlevel floors with loose doors or struggle to burrow her way into her own home on the top floors. If you don't wish to move in with other queens already inside, you must spend an hour or so digging your own hole, under the eyes of lizards and birds lurking around for food.

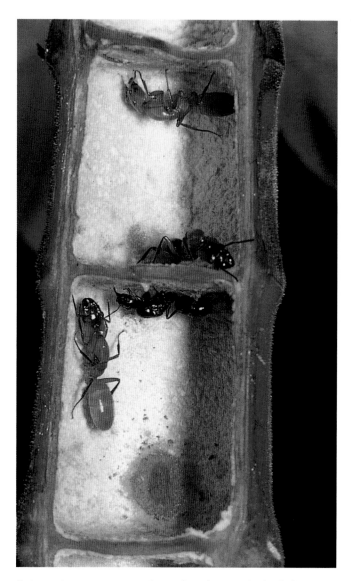

Aztec ant queens cooperating to found new colonies in inter-
nodes of a trumpet tree. The upper node contains queens from
the same species making a nest together; the lower node contains queens from different species founding a colony—an ant
international alliance.

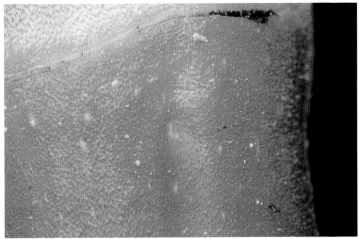

A queen ant burrows her way into a trumpet tree.

Opposite, An Aztec ant queen lands on a trumpet tree, looking for her new home (*top*); the thinner parts of the trunk are easier for the queen ant to burrow into (*bottom*).

Tagging the Queens and Observing Them by Endoscope

One day Dan came up with an ingenious idea of using a medical endoscope to observe ants inside a trumpet tree. In the past there had been no way to know how a queen ant lived once she was locked in. Most human apartment buildings have large windows that give us a great view of the people living inside them, but it is not so easy to observe the inside of a tree without killing it. Previously, Aztec ant queens were observed by splitting open the trees that they lived in, catching them, and raising them in laboratories. This cannot compare to observing them in their natural state.

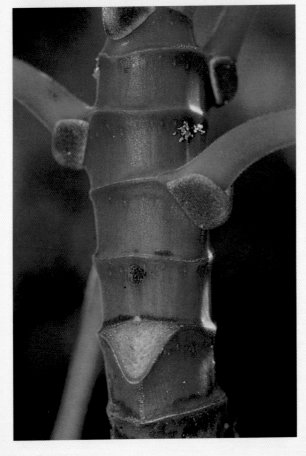

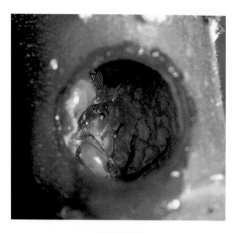

Top, A hole we drilled into a live trumpet tree trunk to observe the inside. *Bottom,* We tagged each ant queen we observed.

Opposite, A queen ant scrapes off part of the inner wall of the tree trunk to plug the hole she made when burrowing in (*top*). Trumpet tree—the Aztec ant queens' high-rise apartment building (*bottom*). The bottom floors are closed to newly arriving queen ants, but home-seekers can move into places that are easier to burrow into with roommates—such as the third floor from the top or the top floor.

Drilling a peephole is simple enough, but observing the day-to-day activities of ant queens in the pitch-black darkness inside a tree trunk is another matter altogether. An endoscope has a long tube with a small lens on the end surrounded by a non-heat-producing fiberoptic light source, designed to look inside small, dark spaces.

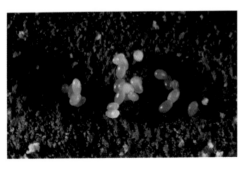

In the days when we did this research, medical doctors were just beginning to use them to look inside people's intestines and to perform knee surgery. Deep in the tropical rainforest, away from human civilization, we used this cutting-edge medical technology to be the first human beings to spy on this Aztec civilization in situ.

Aztec ant queens may live together, but this certainly does not necessarily mean that they cooperate with each other. We still needed to determine whether the ants that lived together all participated in laying eggs and worked together to raise their young. We could see that all of the queen ants that lived together had developed swollen abdomens. This proved that they were all carrying eggs. The fact that they were all carrying eggs did not guarantee, however, that they actually laid and hatched all of these eggs.

We could determine which queens were laying eggs and which ones were not by observing them round the clock, but this was easier said than done. Not only could this lead to excessive interference with the ants' lives but it was also undesirable for us. Dan came up with another ingenious idea. He injected a different-colored dye into each queen to find out which queen ants were laying which eggs. The results were dazzling. Each queen laid eggs of a different color, depending on which dye had been injected into her. This proved conclusively that Aztec ant queens do cooperate with one another.*

*D. L. Perlman, "Colony Founding among Azteca Ants," PhD dissertation, Harvard University, 1992.

Opposite, My colleague, Dan Perlman, using an endoscope to observe Aztec ants inside a trumpet tree (*top*). Four queen ants that formed a single colony together (*middle*). Each of the four displays an abdomen swollen with eggs. The queen ants' food was dyed, each queen given a different color so that eggs would be color coded (*bottom*).

Detailed Behavioral Observations
in the Laboratory

After we completed our rough observations of the Aztec ant queens' lives in the field, we still needed to take them to our laboratory to observe them more closely. Ants have no qualms about doing all of the things that they would do in the wild when they are in a laboratory, as long as they are provided with an adequate environment. I still remember the day I took 400 Aztec ant queens that I had collected in the field to our laboratory at Harvard. Memory of the look given me by a quarantine officer at Miami Airport when I told her one of my suitcases was full of queen ants makes me smile to this day. I didn't have to open my bag for her.*

*Jae C. Choe and Dan L. Perlman, "Social Conflict and Cooperation among Founding Queens in Ants (*Hymenoptera: Formicidae*)," in Jae C. Choe and Bernard J. Crespi, eds., *The Evolution of Social Behavior in Insects and Arachnids* (Cambridge: Cambridge University Press, 1997).

16

War and Slavery

*from full-scale slaughter
to bloodless warfare*

A soldier's severed head rolls onto the ground. There have been heavy losses all day in this seemingly endless battle. Every soldier on the field faces impossible odds, outnumbered by as many as two or three to one. There is no way they can possibly stem the tide of the enemy's relentless assault. Some of the swifter soldiers race to the rear to call in reinforcements. . . . A battle scene that sounds as though it could have come from a war in ancient mythology. But this war story comes from today's real-life battlefields of the ants.

While we do not yet know if there are frontline commanders who shout commands like "charge" or "retreat," scientists have observed messenger ants that race back to the nest when reinforcements need to be called in. Bert Hölldobler discovered this phenomenon during more than a decade of observing the battlefields of the honeypot ants of the Arizona desert.

Ascertaining the Strength of the Enemy to Formulate Strategy

Ants have no concept of numbers as we do, so how can they estimate the strength of their enemies' forces? They definitely do not do it by counting the number of soldiers on the enemy's side and then comparing that number to the size of their own forces. Human beings are believed to grasp the concept of numbers at age five or six, approximately. Studies have also shown that some animals, such as some birds, mice, and nonhuman primates, also understand the elementary concept of numbers. Crows and parrots demonstrated

A large-bodied worker and a small-bodied worker of the Costa Rican *Pheidole* ant. The more large-bodied workers in a colony the stronger is its army.

their ability to make the connection between numbers and specific amounts of objects quite some time ago. A chimpanzee named Sheba has even been observed performing single-digit addition.

Ants seem not to be able to count objects per se, but biologists speculate that ants can probably indirectly compare the strength of their own forces to that of their opponents. Bert Hölldobler has observed three ways that ants do this. The first is that they estimate which side is larger based on whether they bump into their comrades or enemies more frequently when they engage each other on the battlefield. If they run into enemy soldiers more often, then it is time to prepare to retreat. When they are doing battle with an ant species that contains different body-sized workers, they may also measure the size of enemy forces based on how often they encounter large-bodied enemy soldiers. Only the largest colonies are capable of producing large numbers of large-bodied workers. This means that when a colony sends a certain number of large-bodied workers into battle, it is also possible to guess how many of the smaller-bodied workers have been left behind, as well as the total size of the colony. The third way involves observing how many enemy soldiers have shown up on the battlefield but are not participating in combat. According to Hölldobler, the honeypot ants use all three.

Bloodshed is common among the savage beasts of the animal kingdom, but not many animals are inclined to fight each other to the death. Animals that engage in large-scale massacres the way humans do are even rarer. The only animals that can compete with human beings in terms of bloody warfare and slaughter are bees, wasps, and ants. Coincidentally, all these animals have developed highly organized societies. Perhaps massacre and tribal warfare are an inevitable side effect of social evolution.

Ant warfare is not always bloody. Ants often use rituals that involve minimal bloodshed and end confrontations relatively peacefully by estimating the strength of each other's colonies. These rituals are much like the Maring tribe of hunter-gatherers' "nothing wars" on the island of New Guinea. In these rituals, the tribesmen don traditional costumes and march single file to their border to shout at their enemies and dance. This ritual is nothing more than a show meant to shock and awe. If no end is forthcoming, the tribesmen will eventually break out their arrows and begin shooting, but as soon as anyone from either side is killed or wounded the battle is over. Full-scale war almost never breaks out.

The Motive Is the Economy, Not Genocide

One reason why honeypot ants rarely engage in bloody warfare is that for them warfare is mostly just part of their economic strategy. The food they prefer is not always easy to find in the desert environment they live in. Once they find a food source, they must harvest it and carry it back to their nest quickly, before neighboring colonies catch wind of what they have found. If they do not, they can expect hostilities to break out and the food to come under siege. To prevent this, the honeypot ants dispatch soldiers to the borders of their neighboring colonies to distract them by engaging them in a contest of strength, while other workers from their colony harvest the food and carry it back to the nest. This warlike display is actually a ruse to help them maintain their economy.

The wily honeypot ants, however, have a formidable rival in their neighbors, the *Conomyrma bicolor* ants. If the *Conomyrma bicolor* ants find a good source of food first, they immediately send troops to the honeypot ants' nests and surround the entrances. Once they have laid siege to the honeypot ants' nest, they carry small rocks in their jaws and drop them into the entrance. While the honeypot ants are preoccupied with this landslide of rocks, other *Conomyrma bicolor* workers harvest the food. Even the most skilled tacticians meet their match.

Conomyrma bicolor ants using tactics of distraction. When they discover a food source, they send troops to the nearby honeypot ants' tunnel to drop rocks down the entrance. When the honeypot ants get busy clearing the stones out of their nest, they do not notice that the *Conomyrma bicolor* ants are taking their food.

While humans engage in brutal conflicts over differences in ideology or religion, ants fight wars primarily for economic reasons. Human beings, too, may go to war for economic reasons, but those wars are nowhere near as cruel as the wars they fight over their beliefs. As we can see from the atrocities of the Nazis during the Second World War and from more recent conflicts, in Bosnia and Rwanda, for example, humans at war will not hesitate to engage in genocide in the name of their beliefs. Humans are the only animal that has "got religion." It may be fortunate that no other animals have discovered a taste for it.

Ants and Humans: The Only Ones to Practice Slavery

One of the main reasons ants go to war is to capture slaves. When honeypot ants discover through their ritual tournaments that the opposing colony is particularly weak, they immediately invade the weaker colony's nest, confiscate honeypots, and kidnap that colony's younger workers and larvae to enslave them. Honeypot ants are one of the few ant species that enslaves other ants of their own species; most slave-raiding ants abduct the larvae of other species. To look at this "slavery" in the context of human behavior, the honeypot ants' abduction tactics can be said to be slavery as we know it, whereas the other ants' use of other species of ants could be more accurately compared with our use of horses and oxen as livestock.

Amazon ants of the genus *Polyergus* are the best-known genus of slave-raiding ants. To see them invade the nests of other ants to capture slaves is a remarkable sight. Once they choose a colony to pillage, they send a few scouts to reconnoiter the nest and leave an odor trail on their way out. The rest of the raiders follow the odor trail, charging into the nest at full tilt. Multiple columns sprinting side by side dash into the nest at the speed of three centimeters per second. Proportionately, adjusting for size, this is the equivalent of a column of human soldiers in full battle gear marching at 26 kilometers per hour (16 mph).

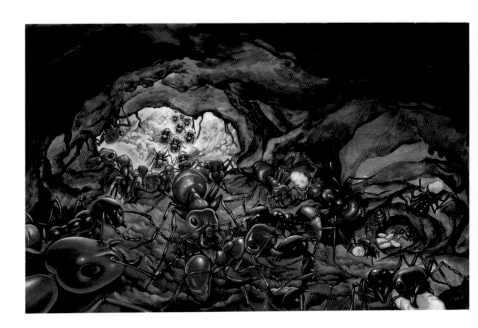

Top, The red *Polyergus* slave-raiding ants are invading another species' nest to abduct their larvae to use as slaves. The black ants are rushing to find a safe place hide their larvae from the intruders. *Right*, This cartoon depicts a slave-raiding ant enslaving another ant.

Polyergus slave-raiding tactics vary by species. Some species follow their scouts directly into the target nests, while others follow odor trails left by scouts. They kill any enemy soldiers they see, using their scimitar-like jaws, and penetrate deep into the enemy nest, triumphantly returning to their own nest with any larvae that they find. When these larvae hatch into adult ants, they are bathed in the pheromones of their new queen, and they will spend the rest of their lives serving the colony of their kidnappers. It is as if they have been brainwashed. Once they have captured their slaves, the slave-driving worker ants do virtually no work aside from overseeing their slaves, which leaves the masters free to loaf around the nest. In experiments where the slaves have been removed, the slave-driving worker ants, except the ones with heavily modified body parts, show themselves to be perfectly capable of doing all of the colony's work on their own. Once their slaves have been returned, however, they immediately stop working and go back to relying on their slaves.

Most species of slave-raiding ants, however, do not capture their slaves by overpowering and slaughtering enemy soldiers the way the Amazon ants do. They use much simpler means, and sometimes do so without bloodshed. For example, more than half of the space in the *Formica subintegra* worker ants' abdomens is taken up by Dufour's glands, which they use to secrete a kind of "propaganda pheromone" that instills chaos in the nests they raid while they capture larvae. The larvae then become slaves. In the case of the *Epimyrma* and *Harpagoxenus* ants, once a young queen has completed her nuptial flight, she finds a colony of *Leptothorax* ants to infiltrate, assassinates their queen, and assumes her throne.

They Are Everywhere We Look

If you look around you, it probably will not be hard to observe ant warfare where you live. On a warm spring day, while the cherry trees next to my laboratory were in full bloom, my research team discovered and observed a war that lasted several days between Japanese carpenter ants living underneath the trees. We never found out exactly what they were fighting over, but judging by the number of soldiers sent into battle, it must have been important to them. One worker faced off against an enemy soldier, holding back, waiting for her chance to immobilize her opponent by grabbing a leg or an antenna so that her comrade waiting in the wings could bite off the enemy's head or abdomen. It was not clear whether these Japanese carpenter ants sent messengers to the rear to call in reinforcements when their forces were weak, the way the

honeypot ants do, but the number of soldiers that each side sent to the battlefield appeared to be one of the key factors in determining which side emerged victorious.

There were many casualties in this protracted battle. One worker ant kept fighting the good fight even though her abdomen had been completely sliced off. Another ant continued to engage the enemy with the head of an enemy soldier dangling from her antenna. The head must have belonged to an enemy soldier who had grabbed the antenna only to make herself vulnerable to attack; she ended up decapitated.

Opposite, A worker ant faces her enemy after her abdomen has been severed off (*top*). A worker ant rushes into battle with an enemy soldier's severed head hanging from her antenna (*bottom*). © Hangkyo Lim

Epilogue

. . . .

to know them is to love them

The Chinese character for ant (蟻) is a combination of the characters for righteousness (義) and insect (蟲). Thousands of years ago, the Chinese had already understood the unusual ability of these righteous insects to sacrifice their individual needs for the good of their society.

Ants are not particularly gorgeous animals. Even in the most amazing of the photographs in this book, the ants themselves are rarely as attractive as their surroundings; in fact, making a book based around photographs of ants is like making a movie mainly with supporting actors. The stars shine less brightly than their background.

The charm of ants, however, has little to do with the way they look. It is their mental world that allows them to rival humanity in their ability to build organized societies. At first glance, ants appear to hold more totalitarian political ideals than we do; nevertheless, our democracies can learn a lot from them. In a democracy, although we must respect the rights of the individual we must also pursue the collective welfare and the development of the nation.

Human existence spans a mere 4 to 6 million years, while ants have had at least 80 million years to experiment with their social structures through trial and error. If we take the French Revolution as the origin of democracy, this human social experiment has lasted little more than two hundred years. Marx's ideology may have failed, but it would have meaning to ants. As Edward O. Wilson once said, "Karl Marx was right, socialism works, it is just that he had the wrong species."

Human beings often consider themselves to be the masters of all creation. Let us suppose, however, that biologists from another planet came to Earth to study the creatures that live here and the ecosystems they inhabit. These aliens would certainly be interested in taking detailed notes on the Earth creatures that have so actively altered their environment by building cities, with their skyscrapers and spiders' webs of roads, and farms that grow such a wide variety of crops. They would probably be surprised to learn that just one species, *Homo sapiens,* has managed to accomplish all this. There is no question that human beings are the masters of this remarkable, mechanized modern civilization. But we must also accept an undeniable fact: lowly insects control most of the territory on this planet.

Human beings may dwarf ants in physical size, but ants control at least as great an amount of the Earth's ecosystem. Ants and other social insects have achieved an amazing level of ecological success. The only places on Earth that ants do not live are the peaks of the highest mountains, the polar regions, and under water. We can find ants living deep in the densest forests, in deserts where it is hard to find so much as a blade of grass, inside human homes, and on the sidewalks of crowded cities.

In the mid-1970s, German ecologists estimated the weights of every animal living in the Amazon rainforests. Then projecting on a large scale the weight of the samples they had taken, they determined each animal group's approximate total biomass. Surprisingly, they determined that even though on the individual level ants and termites are less than one-millionth the size of human beings, they make up nearly one-third of the total animal biomass of the Amazon rainforests. A single ant may not seem like a force to be reckoned with, but add them together and in terms of sheer mass they beat out larger animals like jaguars and tapirs for the title of dominant species, rulers of the jungle.

According to the latest estimates, approximately 12,000 species of ants live on this planet. When every ant is counted, some say that there may well be as many as 90,000 species. Although ants represent but a tiny fraction of the total number of insect species, which could be as many as 30 million, in terms of biomass they are overwhelmingly among the most dominant insects. Guess how many ants are living on Earth. Most ant biologists agree that there must be at least 100 trillion (10^{14}), or even as many as 10 quadrillion (10^{16}) ants out there. If we count the mass of the average worker ant at between 1 and 5 milligrams, the weight of ants would overwhelmingly outweigh that of the entire human population—soon to be 7 billion (7×10^8).

Ants and humans may have around the same total biomass, but because ants are much smaller there are few ecosystems on Earth that they have not made a niche in. Many species of ants are predators that hunt for other insects and small animals, but there are also scavenger species that feed on the rotting carcasses of already-dead animals. Scavenger ants feed on and clear away nearly 90 percent of all animal carcasses, making them nature's number one garbage collectors.

A remarkable number of plants cannot reproduce without ants' assistance. Earthworms are well known for tilling the soil, but in half of Earth's ecosystems, particularly in tropical regions, no animal serves the planet better than ants in terms of turning and recycling soil nutrients.

Ants have been on this planet far longer than we have and they will likely still be here long after we are gone. Were the human race to become extinct, what would happen to the world's ecosystem? Alan Weisman painted a puckish but morbidly imaginative future in *The World without Us.* The ruins of the many things we have built would remain for quite some time after we have passed, but that would have no great impact on the ecosystem. Things made of bronze might survive for millions of years, and more than a billion tons of man-made plastics might remain for quite some time. Given thousands of years, however, polymer-eating microbes might evolve, and Earth might revert to Eden. Human beings are the greatest predators on Earth in terms of the animal bodies we have put on our breakfast tables alone. Likewise, we have mercilessly ruined the habitats of countless animals in order to maintain our lives of ease and comfort. If we disappear, the ecosystem will undergo significant changes in form and function. These changes, however, on the whole can only be good for the animal kingdom. If human beings disappear, virtually no animal, besides perhaps cockroaches and rats, possibly cats and dogs, would mourn our loss. Were ants to disappear, however, the ensuing changes would threaten the very existence of the Earth's ecosystem.

Although ants are tough creatures that can adapt relatively well to damaged environments, if the plants and animals they depend on disappear, the ants, too, inevitably will disappear. Our environmental problems are approaching the point of no return. If we keep putting off solving these problems in order to keep trying to live better, we will fall into an abyss from which there is no escape. Cleaning up and conserving the environment is not something we can put off until tomorrow. We must do it now. This is not the job of a future generation—it is *our* job.

I enjoy visiting kindergartens and elementary schools to talk to children about animals. I particularly enjoy talking to them about ants. One day my son's kindergarten teacher asked me to show her five year olds and six year olds slides of ants and to tell the children about the ants in the slides. The children clamored to ask me questions. After the class was over, I set up an ant farm in their classroom populated with ants that my son and I collected from the mountain behind our home. According to a letter I received from the teacher, her students watch the ants in the ant farm for hours every day. When they go outside to play, they no longer step on the ants they find. Instead they lie on their stomachs to watch the ants lay their odor trails. The teacher thanked me for teaching her students to love nature.

In the words of the English philosopher Sir Francis Bacon, "knowledge is power." But knowledge is also love, and once you get to know ants, you'll love them. People have to know each other to love each other. If we do not know each other well, we misunderstand and hate each other. When we truly understand one another, we cannot help but love each other. The same can be said of nature. Once we come to love nature, we will not be able to harm it, no matter how much others might push us to do so. To love nature is a far more effective basis for conservation education than relying on our schools to require children to pick up trash from streams. The best thing we can do for nature is to study it, love it, and then use what we learn as the basis for educating the public and the next generation. The "ignorance is bliss" aphorism absolutely does not apply to conservation. I will never forget the children who screamed in fear when I began my slideshow about ants, but then after the lesson volunteered to help me set up the ant farm.

In Korea, nearly 135 different species of ants have been described. This is a surprisingly large number for a small, peninsular country. It is even more surprising that the figure comes mostly from the better-known fauna in the south of the peninsula— South Korea. That is approximately three times as many species as in Great Britain or in Finland. In the United States, the state of Louisiana, which has about 30% more surface area than South Korea, has about 130 species of ants. But these numbers cannot begin to compare with the diversity of ants in the tropics. A single eight-square-meters patch of the Peruvian rainforest alone is reported to be inhabited by more than 300 different species of ants. Scientists found a single tree that housed 43 different species of ants, rivaling the totals for Finland or Great Britain.

Because ants are so adaptable, they can be found everywhere except for the highest mountain peaks. It is difficult to find a place that ants do not inhabit, from mountains and fields to a busy downtown and inside people's apartments. Anyone can easily observe ants in their natural environment and even perform simple experiments involving ants.

Lately, pharaoh ants have been in the news. People in upscale high-rise apartment complexes have complained about being plagued by the comings and goings of the tiny pharaoh ants, named for the ancient Egyptian monarchs. These are no ordinary ants. In almost all other ant species, the young queen finds males from a different colony during her nuptial flight and founds a new colony after mating; that is how ant colonies usually spread. Pharaoh ant queens, however, mate with males from their own colonies—with their brothers.

Pharaoh ant colonies are often ruled by many queens. There are pharaoh ant colonies on record that contain as many as 2,000 queens. This type of polygynous, or multiple-queen, ant species is unusual in the ant world. Just as all human nations seem originally to have been governed by individual, absolute rulers, it is also common practice for ant nations to be governed by a single queen. Pharaoh ant colonies are an exception. There is an old saying that too many cooks spoil the broth, meaning that if too many people are in charge it is hard to govern. In addition, the frequent incestuous mating that occurs among the pharaoh ants may cause their gene pool serious problems. Still, it is not hard to see the advantages of the pharaoh ants' way of life. Once their ant queens have finished mating, they already have everything they need to start a new colony in a new location. This is why pharaoh ant queens have been able to pack themselves away so easily in travelers' baggage or in cargo to spread their families from their native Africa to countries all over the world. I would bet that the number of countries invaded by pharaoh ants is greater than the entire membership of the United Nations. Pharaoh ants have achieved this amazing success by violating the social taboo of incest.

There are pharaoh ants living in my home. They do not bother us, my son and I; we treat them like welcome guests. In fact, we are grateful that this bit of nature has come into our home in the middle of the city. Some time ago, the producers of a cable TV show that was doing a special on pharaoh ants came to my lab to ask me about them. They sent a team to film footage of the ants I am raising in my lab and to interview me, but when I told them that I had no intention of helping them if their goal was to teach people how to kill ants, they looked perplexed. An ant infestation is almost impossible to eradicate, unlike cockroach infestations. There is actually a brighter side: ants will actually drive other, more harmful insects like cockroaches out of your home. There is no need to kill them. Furthermore, ants in the home provide a golden opportunity to learn more about nature by observing the ants, as my son and I do. I convinced the TV people to change the slant of their program and agreed to participate in the broadcast. I heard later that the viewers' reaction to the program was positive.

Perhaps the easiest ant behavior to observe is how they find food and carry it back to their nests. When they find large sources of food, it is particularly interesting to watch how they communicate this information to other ants in their colony. As we saw earlier, there are ant species whose workers drag their fellow workers, one on one, to better sources of food, but most species use chemical means of mass communication to relay this kind of information to the rest of the colony. It is easy to spot them dragging the tips of their abdomens along the ground to leave an odor trail. If you like the idea of observing ants but really don't like hundreds of little pharaoh ants crawling all over your home, try to find which route they use to enter your home, then leave some goodies not too far from that entrance everyday. Ants are so economically rational that they will not waste their time roaming around elsewhere to look for other sources of food; instead, they will be fully occupied harvesting what you provided for them. If you are willing to share a tiny corner of your home, you can enjoy nature right in your living room.

It is also fairly easy to find ants tending to and collecting honeydew from aphids and scale insects. Look carefully enough and you can find them even in modern apartment complexes. There are even ants that collect honeydew directly from plants. In addition to the many plants with nectaries inside their flowers, many plants have extrafloral nectaries, and these are almost always for the benefit of ants. These plants provide honeydew to ants, and in return the ants protect them from the various herbivores. As I write, next to my laboratory building I am observing cherry trees with ants that drink from their extrafloral nectaries.

Another exciting phenomenon to observe is a queen ant's nuptial flight. The best times to observe queen ants taking off to start their new families in most temperate regions are the warm, breezy days of May and June; September and October are also good times.

The hundreds of queen ants in one place flying higgledy-piggledy to meet their many sperm donors is an amazing sight. The best place to observe this event is on top of a hill—a favorite place for queens to go for their rendezvous. This may be because "meet me in the highest place in the area" is a simple message to share from generation to generation. It may not be the most romantic place to meet, but there is less room for misunderstanding.

If you catch an ant queen that has just broken off its wings and is digging a hole for its new home, then raise it in an ant farm and you will be able to observe how a queen founds a new queendom. It can also be worthwhile to catch an ant queen that has already founded a colony. It should be a relatively simple matter to find ant colonies under rocks or inside a rotten log. If you intend to keep an ant farm for any length of time, you must capture a queen. It is easy enough to dig up an ant colony, but catching a queen is not so simple. It takes patience and tenacity. To learn more about capturing and raising ants in captivity, refer to the appendix "How to Study Ants" in *Journey to the Ants* by Bert Hölldobler and E. O. Wilson.

The stories in this book barely scratch the surface of the amazing world of ants. It is my hope that this book inspires many young readers to join the ranks of the ant researchers of the world, as amateurs or professionals. The ants need us to study them, so get to know them better and love them.